AA003061

North American Mixing Forum 2014

Core Programming Area at the 2014 AIChE Annual Meeting

Atlanta, Georgia, USA
16-21 November 2014

ISBN: 978-1-5108-1266-6

Printed from e-media with permission by:

Curran Associates, Inc.
57 Morehouse Lane
Red Hook, NY 12571

CURRAN ASSOCIATES INC.
proceedings
.com

Some format issues inherent in the e-media version may also appear in this print version.

Printed by Curran Associates, Inc. (2015)

AIChE
120 Wall Street, FL 23
New York, NY 10005-4020

Phone: (800) 242-4363
Fax: (203) 775-5177

www.aiche.org

Additional copies of this publication are available from:

Curran Associates, Inc.
57 Morehouse Lane
Red Hook, NY 12571 USA
Phone: 845-758-0400
Fax: 845-758-2634
Email: curran@proceedings.com
Web: www.proceedings.com

TABLE OF CONTENTS

394101 Lattice-Boltzmann-Based Algorithms for Predicting Impeller Performance

Abstract:

A novel approach for performing large-eddy simulations (LES) of industrial-sized mixing processes is presented. This approach, which is an extension of the lattice-Boltzmann model (LBM), is trivially parallelizable and runs such simulations orders-of-magnitude faster than conventional CFD tools. The mixing properties of a Rushton turbine, as predicted from the LBM model, are compared to experimental data. The vortex structure, eddy-dissipation rates, pumping number, and power number predicted from the model agree with experimental data across multiple Reynolds numbers. Since fluid-structure interactions are described using an immersed boundary method, system set-up is trivial and no explicit meshing process is required. As such, simulations involving more sophisticated impeller morphologies present no meaningful increase in system set-up and runtime. Operating on modern computer architectures (e.g., GPUs, Xeon Phi, etc.), the model can simulate multiple seconds of a mixing per hour of processing time.

374885 Design of Small Scale Impellers for High Throughput Mixing Challenges

Abstract:

With the increasing use of automated high throughput research labs to perform rapid screening experiments, we are often faced with unique mixing challenges on a fairly small scale. In this talk we will present some examples of small scale impellers that have been developed to meet specific goals in high throughput equipment where the mixing vessels are small unbaffled tubes or vials. We have employed computational fluid dynamics and rapid prototyping capabilities to enhance our ability to analyze and optimize the impeller geometry.

One example will focus on a parallel reactor system where one of the reactants is contained in the head space gas. The mixing impeller has to effectively draw down the gaseous reactant into the liquid and provide adequate fluid and thermal mixing in the tube. Due to the manner in which the drive mechanisms is set-up, the impeller needs to perform in the same manner whether it is turning clockwise or counterclockwise. It also needs to be inexpensive to manufacture since it is disposed of at the end of an experimental run. This led to a 2D impeller design that provided an efficient compromise between draw down and axial flow development.

In another example involving a multi-step robotic system, it is imperative that the impeller does not draw down any air bubbles into the fairly viscous fluid during the blending step. The impeller torque also needs to serve as a viscosity indicator that can be calibrated against a lab bench viscosity analyzer. In this case, since the impeller is re-used, the design needs to allow for easy and effective cleaning in a separate robotic step. For chemical and mechanical integrity the impeller needs to be constructed of metal, but once again a low impeller cost is a primary goal.

In a third example, a unique impeller based on a twisted tape geometry was developed for rapid blending of high viscosity fluids in the range of 5000 to 30,000 cp. A number of geometric parameters were evaluated and optimized. The impeller is able to generate good axial mixing and is a significant improvement over the performance of the small scale impellers offered by venders.

377172 Vortical Flow in Rotating Milli- and Micro-Chambers for Mixing and Process Intensification

Abstract:

A novel method using transient-generated inertial effect is used to mix a small sample (milli- to micro-liter) in a chamber in a rotating Compact Disk. The primary vortical flow generated from inertia further induces a complex Coriolis vortical flow, which in conjunction with the primary flow, speeds up mixing by orders of magnitude in an otherwise viscous-force dominant small chemical reactor.

Experiments and numerical simulations are used to reveal the complex three-dimensional spiral toroidal flows generated and associated mixing. Further, the scale-up laws have been developed for process design. Such process intensification can be applied to cell culture, chemical reaction studies, and many more applications.

Shaking bioreactors are widely applied for screening and bioprocess development projects, due to its flexibility and ease of operation. It has been assumed that probably more than 90% of all culture experiments in biotechnology are performed in shaking bioreactors, in which orbital is the main shaking motion [1].

A major deficiency in using flasks as shake culture reactors is that the oxygen transfer rates obtained are comparatively low, and oxygen supply may become limiting if the oxygen uptake rate of culture exceeds the oxygen transfer capacity through the shake flask closure and/or the gas-liquid interface. In that case, the observed growth is a function of the oxygen transfer rate, and it does not reflect the actual growth rate of the microorganisms [1], [2].

The oxygen transfer rate (OTR) from the headspace to the liquid phase is described by equation (1): $dc_L/d_t = OTR = k_La(C^* - C_L)$. Where *OTR is* the change in oxygen concentration over a time period; k_La is the volumetric mass-transfer coefficient; C_L is the concentration of dissolved oxygen in the liquid phase and C^* is the saturated dissolved oxygen concentration. The k_La coefficient is a measure of the gas supply global capacity of any bioreactor. Different correlations have been proposed for k_La estimation in orbital shake flasks [1], [3].

ResonantAcoustic® Mixing (RAM) technology enables noncontact mixing by the application of low frequency acoustic energy applied to a vessel. The chaotic motions created within the mixing vessel cause a uniform distribution of micro-mixing cells as well as creating bulk mixing flow. Applications of these technology include mixing a plurality of fluids, e.g., intimately mixing a gas in a liquid, or a liquid in another liquid, or more than two phases, and dispersion of solids in liquids, in particular hard to wet solids and small particles [4].

Recently, Applikon Biotechnology® has introduced RAM technology within de RAMbio incubator for mixing microbial cultures, and it is proposed like an alternative to solve oxygen transfer limitations in shake flasks when combined with the Oxy-Pump® stopper, which actively pumps air in and out of the vessel.

To get a better understanding of the OTR in acoustic shake flasks, we evaluated the k_La under different mixing conditions, and compare it versus the k_La obtained in orbital shake flasks. Further, we followed biomass growth and dissolved oxygen tension (DOT) during an *Escherichia coli* cultivation in orbital and acoustic shake flasks at equivalent k_La values.

The k_La was evaluated in flasks, orbitally and acoustically shaken, under different conditions of mixing velocities (50-350 rpm and 3-20 g, respectively), flask volume (250, 500 and 1000 ml), filling volume (20 and 40 %), at either 30 or 37°C. The k_La was determinated as the slope of the equation resulting from integration of equation (1). DOT was recorded after the water was deoxygenated (by sulphite oxidation), and the shakers either orbital (New Brunswick Scientific C251 with an orbital diameter of 254 mm) or acoustic (RAMbio) were started. DOT was measured online with a luminescent patch PSt3, on a Fibox3 by PreSens.

Results of k_La measured in orbital shake flasks are comparable with those predicted with the model proposed by Büchs [3]. Out-of-phase phenomena was observed at high agitation rates (over 250 rpm) in 250 ml flasks with 20% of filling volume. Under all conditions tested, the RAM system achieved higher k_La values than those obtained with the orbital shaker.

The k_La observed in acoustic shake flasks are proportionally related to gravity acceleration and to the initial area-volume ratio. Data was fitted to a multiple linear regression with OriginPro 8 software. The model generated (data not shown) had a coefficient of determination (R^2) of 0.948 and 0.945 for k_La values at 30°C and 37°C, respectively. The model proposed was validated with experimental data with an accuracy of ±25%.

In the other hand, k_La values obtained with the orbital shaker did not fit well to a multiple linear regression in function of agitation rate and the initial area-volume ratio. Models generated had a R^2 of 0.785 and 0.891 for k_La at 30°C and 37°C, respectively. This low correlation are surely due to the out-of-phase phenomena at high agitation rates.

On the other hand, *E. coli* 53606 was cultivated in Luria-Bertany broth at 37°C for 8 h, with either orbital or acoustic mixing at initial k_La near values of either 20 or $70h^{-1}$. In those cultures at a k_La near $20h^{-1}$, a slightly higher specific growth rate was found with orbital (0.90±0.04 h^{-1}) than with acoustic mixing (0.81±0.02 h^{-1}), but exponential growth lasted longer and final biomass concentration achieved was higher with acoustic (1.57±0.13 OD_{600nm}) than with orbital mixing (1.31±0.05 OD_{600nm}) . DOT profiles exhibit a longer period of oxygen limitation in orbital (100 min) than with acoustic mixing (70 min).

Similar results were obtained with cultures of *E. coli* grown at k_La near 70 h^{-1}. The growth rate was slightly higher with orbital (0.94±0.01 h^{-1}) than with acoustic mixing (0.72±0.01 h^{-1}), but exponential growth lasted longer and final biomass concentration achieved was higher with acoustic (3.94±0.06 OD_{600nm}) than with orbital mixing (2.27±0.08 OD_{600nm}). Oxygen dissolved profiles exhibit a sooner exhaustion of DOT in orbital (4 h) than with acoustic mixing (6 h).

Besides, we work with the two different systems for shake flasks (orbital and acoustic), at equivalent initial k_La values, the growth of *E. coli* exhibit and improvement in oxygen supply in the RAMbio system. This could be regarded to an increment of the OTR due to a higher DOT by the formation of small droplets.

In the present work we introduce the first approximation to the understanding of the oxygen transfer rate in shake flasks using the ResonantAcoustic® technology.

Acknowledgements. CONACYT-INNOVAPYME 181895, CONACYT 178528, 104951-Z and PAPPIT-UNAM IN-209113, IN-210013

References

[1] J. Büchs, *Biochem. Eng. J.,* 2001,**7**, 2, 91–98.

[2] U. Maier and J. Büchs, *Biochem. Eng. J.,* 2001, **7**, 2, 99–106.

[3] W. Klöckner and J. Büchs, *Trends. Biotechnol.*, 2012, **30**, 6, 307-314.

[4] H. W. Howe *et al.*, United States Patent No.: US 7,866,878 B2, 2011.

Chemineer™
Reliability & Technology in Mixing

NOV Process and Flow Technologies
5870 Poe Avenue
Dayton, OH 45414
Phone: 937-454-3200
Fax: 937-454-3379

Agitator Design at High Solids Loadings

Many solid-liquid agitation operations, such as mining, occur at high solids loadings (greater than thirty percent solids by mass); however, most laboratory and computational studies are performed at low solids loadings (less than ten percent solids by mass). The work to be described in this presentation is aimed at developing a better understanding of solids suspension agitator design at high solids loadings.

It is known that significant capital and operating expense savings can be achieved if some solids are allowed to settle on the tank base (i.e. - if operation at speeds below the just-suspended speed is acceptable). Operation under these conditions is typically characterized in terms of the relative fillet volume, the fraction of solids that are not suspended, as a function of the relative speed, the rotational speed as a fraction of the just-suspended speed. What is not known is if the limited data on relative fillet volume is general enough to use for design purposes. That is, is relative fillet volume dependent on scale, solids loading, baffling, impeller characteristics (type, size, and off-bottom clearance), and solids characteristics (size, shape, and density)? This study will examine these effects and the development of a design procedure will be discussed.

CFD analysis will be done in conjunction with the experimental portion of this study. Most of the CFD work with solids has been done using a low solids loading (<20%), so validation of the models at high solids loading is critical for the further use of CFD as a design tool. A major component of the CFD experiments will be determining criteria that can be examined to ascertain agitator performance for high solids loading. If numerical values (N_{JS} coupled with a solids fraction parameter) can be used as opposed to visual determination (contour plots of volume fraction), it would aid in the long term implementation of CFD design strategies.

Comparison of Solid Suspension Quality in Two Contactors

Mahsa Taghavi, Sirous Ebrahimi, Jafarsadegh Moghaddas
Sahand University of Technology, Tabriz, Iran

Abstract

The process of solid particles suspension in a liquid is considerably important in the chemical industry. The suspension quality of a recently developed centrifugal impeller contactor is studied and compared with a propeller impeller. The full cloud height of suspension is obtained in the centrifugal impeller speed of 300 rpm and in a 30-fold lower power consumption amount with respect to the propeller impeller. This result approves the significantly higher efficiency of this novel impeller contactor in solid suspension operations.

Keywords: Suspension, Cloud height, Power consumption, Centrifugal impeller, Propeller impeller

Introduction

Mechanically agitated solid-liquid contactors are widely used in chemical processes. It has been shown that in the case of merely dispersion of the two phases or mass transfer operations, minimum impeller velocity for complete suspension is applied, but if chemical reactions occur, the selectivity depends on the liquid hydrodynamics and on the solid distribution in the vessel, and so higher stirrer velocities will be required [1]. Designing and utilizing the impellers that produce homogeneous solid distribution in lower impeller speeds will result in saving energy and is more economic [2]. One of the significant design parameters for a multiphase stirred tank reactor is the power input by the agitator, which is affected by the physical properties, operating parameters, and geometrical parameters such as the type of the impeller. Efficiently performing the mixing process with a minimum expense of energy is usually of great interest [3].

There are two kinds of techniques describing the distribution of solids in a stirred tank. The first one gives accurate quantitative data on local concentrations of solids such as conductivity probe, optical probe, sampling, or tomography methods. The second one gives general observations and semi quantitative information such as cloud height method for the distribution of solids. Cloud height observations will help in selecting vessel outlet locations and in estimating how well mixed the vessel is. In solid suspensions, most of the solids are lifted within the fluid even at speeds above the critical speed of N_{js} to a distinct level. The distance from the bottom of the vessel to this level is called the cloud height. The liquid below this height is solid-rich, while above it there is only an occasional visit by a few small solids [4]. Hicks et al. (1993, 1997) [5, 6], Bujalski et al. (1999) [7], Barresi and Baldi (2000) [1], Bittorf and Kresta (2002) [8], and Khazam and Kresta (2009) [9] have reported extensive data on cloud height and solid distribution in stirred tanks equipped with some conventional impellers.

A novel centrifugal impeller bioreactor operating in a gas-liquid system has been reported by Wang and Zhong [10, 11] to exhibit higher liquid lift capacity, shorter mixing time, and lower shear stress which was potentially applicable to shear-sensitive biological systems. The present study investigates the solid-liquid mixing cloud height and power consumption of this novel centrifugal impeller and compares it with a commonly used propeller impeller.

Experimental

The experimental setup used in this study is presented in Figure 1. A cylindrical flat-bottomed stirred tank made of plexiglass with a working volume of 3 *L*, diameter of *T*=0.13m, and aspect ratio of *H/T*=1.65 was adopted. The vessel had three planar partial baffles with the width of *T*/10 and length of *H/4* standing equally spaced in the upper edges of the vessel. A centrifugal impeller connected to a draft tube located at the clearance of *T*/3 from the liquid surface was utilized. Rotation of this impeller creates a negative pressure in the impeller center, draws liquid and solid particles from the vessel bottom through the draft tube, and finally centrifuges them toward the liquid bulk.

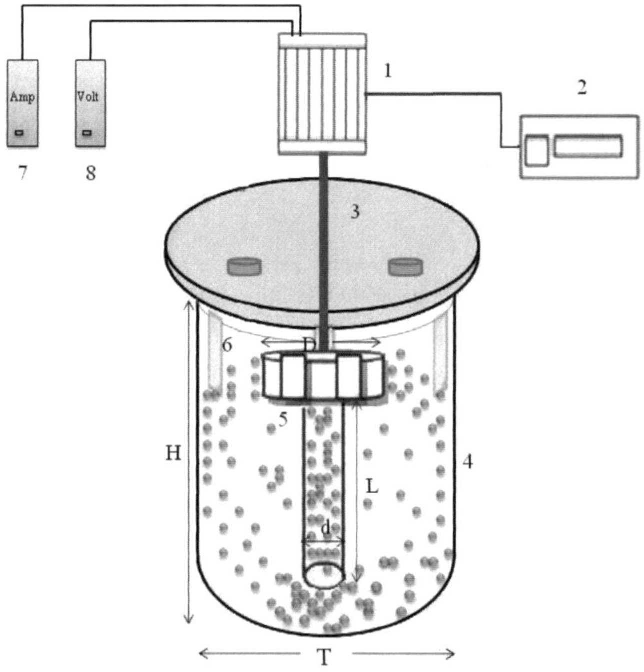

Figure 1. Experimental setup: 1- Electromotor, 2- Stirrer controller, 3- Head plate, 4- Stirred tank, 5- Centrifugal impeller, 6- Baffle, 7- Ampere meter, 8- Voltmeter.

Different agitator designs achieve different degrees of suspensions at similar energy input [4]. For determining the degree efficiency of this recently developed impeller, an axial flow impeller in the type of down pumping propeller (3 segment blade) impeller was used in the same sized reactor with three full length baffles and bottom clearance of *T/4*. The geometrical characteristics of the impellers are given in Table 1.

Table 1. Impellers characteristics

Imp. type	Imp. dia. (*D*)	Draft tube dia. (*d*), height (L)	Blade no.	Blade angle
Centrifugal	7cm	2, 11.3 cm	7	45 °
Propeller	6 cm	-	3	50 °

In this work, the power consumption of the impellers was calculated using the electrical measurement method. In this method the electrical power consumption (P) of the motor is the product of the electrical current (I) and voltage (V) (Eq. 1). For considering the power losses of the system, power consumption was measured in the absence of fluid and then subtracted from the measurements done with the fluid (for more details see Taghavi et al. [3]).

$$P = I.\ V \qquad \text{(Eq. 1)}$$

Cloud height was measured using a ruler and averaged over three runs. Tap water and Ion exchange resins (SG=1.3) were used as the main continuous phase fluid and the solid particles in the fluid, respectively.

Results and Discussion

Cloud height was measured in different impeller speeds for both centrifugal and propeller impellers. As Figure 2 shows, in the impeller rotational speed of 300 rpm, the centrifugal impeller gives a homogeneous suspension and the cloud height (CH) reaches to the fluid height (H) (full cloud height). However, in the case of utilizing the propeller impeller the cloud height reaches to 70% of the total height of the fluid in this speed, and an inhomogeneous suspension is obtained.

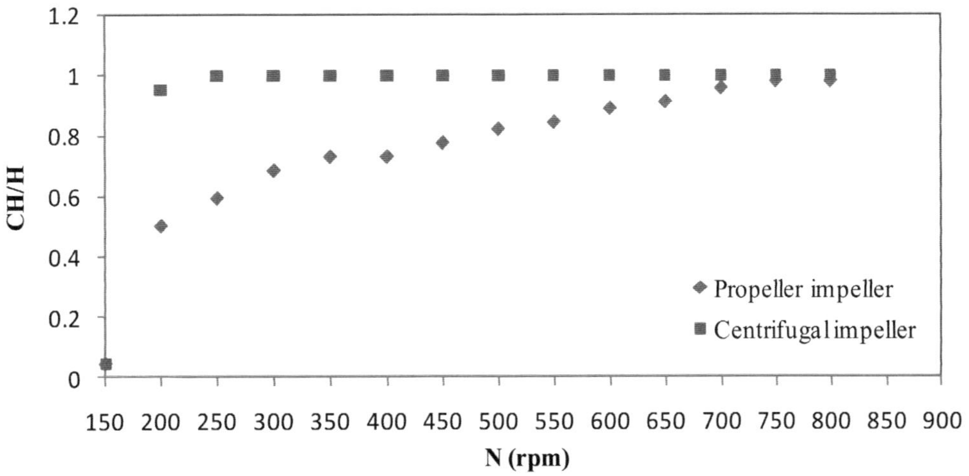

Figure 2. Relative cloud height across the impeller speed

The maximum homogeneity of the propeller impeller occurs in the speed of 700 rpm where CH/H reaches to 95%. Kraume [12] have also reported CH/H=91% in this speed which is in acceptable agreement with the results of this research.

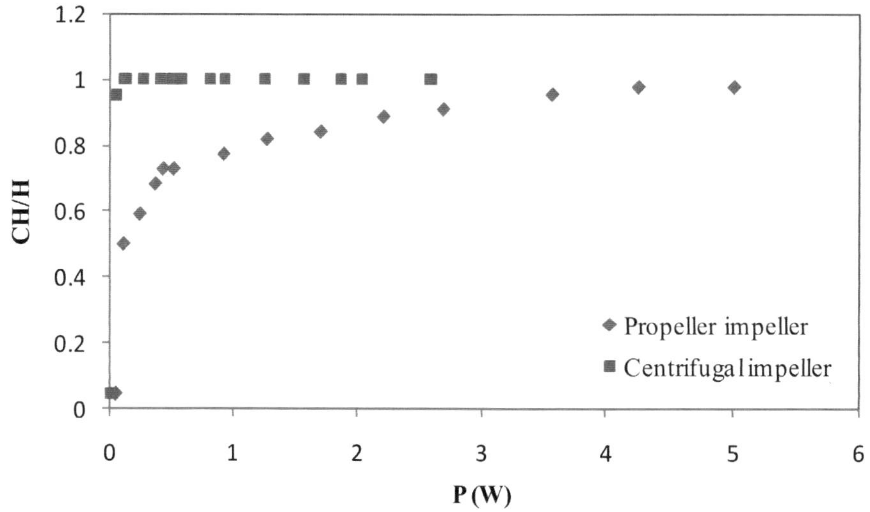

Figure 3. Relative cloud height across the power consumption

Figure 3 demonstrates the variation of relative cloud height with power consumption of the impellers. By utilizing the centrifugal impeller, increasing the power consumption does not affect the amount of relative cloud height, and CH/H=1 could be achieved at low power consumption of P=0.1 W. But, by increasing the power consumption of the propeller impeller, the relative cloud height increases and CH/H=1 is obtained in a 30- fold larger power consumption amount in comparison with the centrifugal impeller.

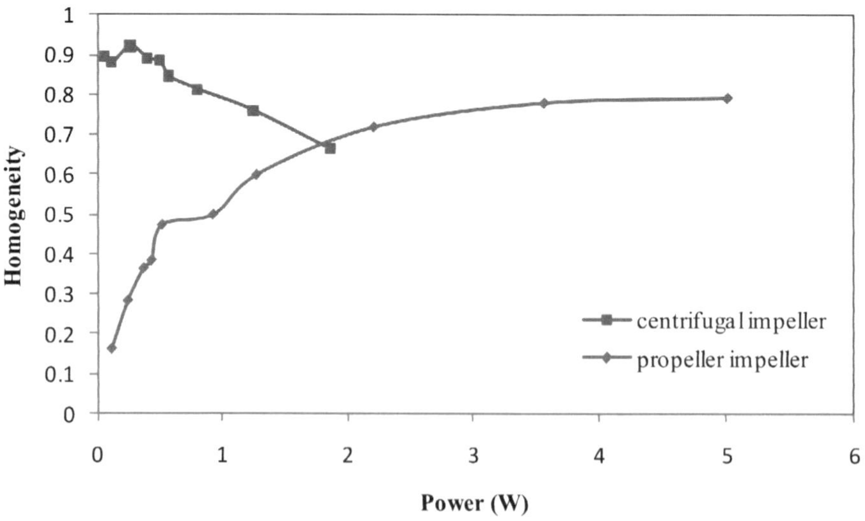

Figure 4. Homogeneity across the power consumtion [2]

These observations are in agreement with local concentration quantitative data obtained by sampling method reported in the previous work by the authors [2]. The higher homogeneity of the centrifugal impeller especially in the low power consumption regions, shows the significant effectivness of this impeller relative to the propeller one (Figure 4).

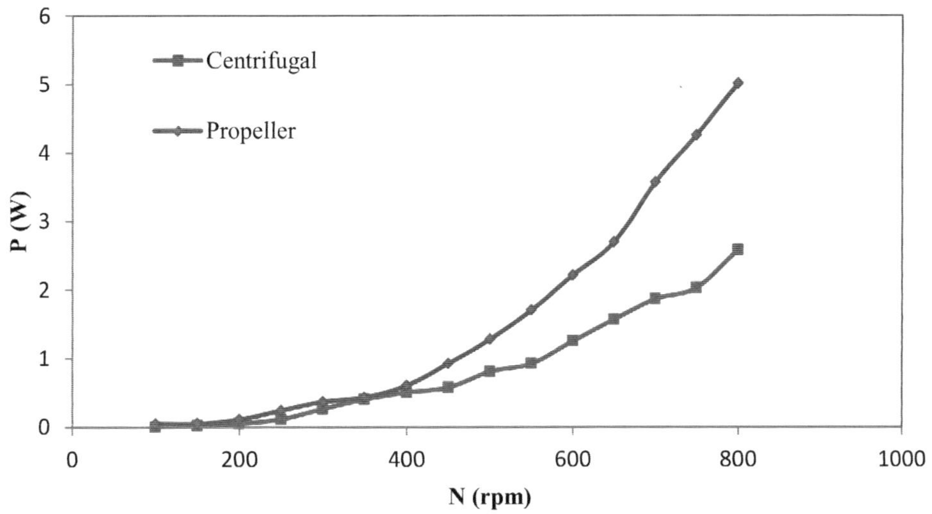

Figure 5. Power consumption across the impeller speed

Power consumption of the impellers have been compared in the Figure 5. It shows that the power consumption of the centrifugal impeller is much less than the propeller impeller especially in higher impeller speeds (N>400 rpm). In low impeller speeds (N<400 rpm) both impellers consume approximately the same amount of power, but as illustrated in Figure 3, in this range of impeller speeds centrifugal impeller produces a complete homogeneous suspension in contrast with the inhomogeneous suspension of the propeller impeller. Thus, the centrifugal impeller is more efficient than the propeller one and produces a homogeneous suspension in lower rotational speeds with very low amount of power consumptions. This figure also demonstrates that increasing the impeller rotational speed obviously causes to increase of the power consumption in both impellers.

Figure 6 compares the variation of power numbers (N_p) with Reynolds numbers (Re). This figure shows the average power number of the propeller impeller is approximately 4-fold higher than the centrifugal one. In the lower regions of Reynolds numbers some fluctuations are observed in the power number values, but by increasing the Reynolds number N_p dose not change and becomes constant. The same trend of behavior has been reported by Bujalski et al. [7] and Sardeshpande et al. [13] for the other types of impellers in solid suspension stirred tanks.

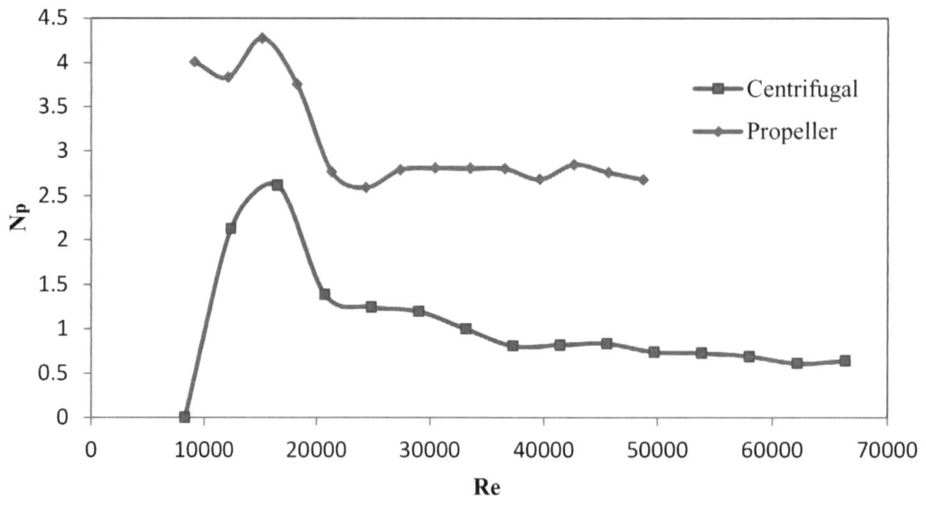

Figure 6. Power number across the Reynolds number

Conclusions

The cloud height and power consumption of a novel centrifugal impeller in solid suspension operation was investigated and compared with a commonly used propeller impeller. It was observed that in the 300 rpm impeller speed, the centrifugal impeller presented a homogeneous suspension and the cloud height reached to the fluid height. However, in the case of utilizing the propeller impeller the cloud height reached to 70% of the total height of the fluid in this speed, and an inhomogeneous suspension was obtained. Moreover, by increasing the power consumption of the propeller impeller, the relative cloud height increased and CH/H=1 was obtained in a 30- fold larger amount of power consumption in comparison with the centrifugal impeller. Finally, it was concluded that the centrifugal impeller was more efficient than the propeller one and produced a homogeneous suspension in lower rotational speeds of the impeller with very low amount of power consumptions.

References

1. A. A. Barresi, G. Baldi (2000), "Power consumption in slurry systems," 10th European Conference on Mixing, Elsevier Science, Amsterdam, pp. 133-140.
2. M. Taghavi, S. Ebrahimi, J. Moghaddas (2014), "Solid Particle Distribution in Centrifugal Impeller Contactors," J. Dispersion Sci. Technol. 35, pp. 1097-1105.
3. M. Taghavi, R. Zadghaffari, J. Moghaddas, Y. Moghaddas (2011), "Experimental and CFD investigation of power consumption in a dual Rushton turbine stirred tank," Chem. Eng. Res. Des. 89, pp. 280-290.
4. E. L. Paul, V.A. Atiemo-Obeng, S.M. Kresta (2004), Handbook of Industrial Mixing: Science and Practice, Wiley-Interscience, NewYork.
5. M.T. Hicks, K.J. Myers, R.R. Corpstein, A. Bakker, J.B. Fasano (1993), "Cloud height, fillet volume, and the effect of multiple impellers in solid suspension," Mixing XIV, Santa Barbara, CA, pp. 20-25.
6. M.T. Hicks, K.J. Myers, A. Bakker (1997), "Cloud height in solids suspension agitation," Chem. Eng. Comm. 160, pp. 137-155.

7. W. Bujalski, K. Takenaka, S. Paoleni, M. Jahoda, A. Paglianti, K. Takahashi, A.W. Nienow, A.W. Etchells (1999), "Suspension and Liquid Homogenization in High Solids Concentration Stirred Chemical Reactors," Chem. Eng. Res. Des. 77, pp. 241-247.
8. K.J. Bittorf, S.M. Kresta (2002), "Prediction of cloud height for solid suspension in stirred tanks," CHISA Conference, Prague, pp. 25-29.
9. O. Khazam, S.M. Kresta (2009), "A novel geometry for solids drawdown in stirred tanks," Chem. Eng. Res. Des. 87, pp. 280-290.
10. S.J. Wang, J.J. Zhong (1996a), "A novel centrifugal impeller bioreactor. I. Fluid circulation, mixing, and liquid velocity profiles," Biotechnol. Bioeng. 51, pp. 511-519.
11. S.J. Wang, J.J. Zhong (1996b), "A novel centrifugal impeller bioreactor. II. Oxygen transfer and power consumption," Biotechnol. Bioeng. 51, pp. 520-527.
12. M. Kraume (1992), "Mixing times in stirred suspensions," Chem. Eng. Technol. 15, pp. 313-318.
13. M.V. Sardeshpande, A.R. Sagi, V.A. Juvekar, V.V. Ranade (2009), "Solid Suspension and Liquid Phase Mixing in Solid-Liquid Stirred Tanks," Ind. Eng. Chem. Res. 48, pp. 9713- 9722.

377549 Solid Suspension in a Stirred Vessel Filled at Different Liquid Levels

Abstract:

Significant hydrodynamic changes have been recently shown to occur in a mixing vessel when the liquid level in the vessel is lowered and the level of liquid submergence of the impeller changes as a result (Motamedvaziri, S. and Armenante, P. M., *Chem. Eng. Sci.*, 2012, **81**: 231-250). These changes have a marked impact on a number of mixing phenomena, including surface air entrainment, mixing time, and solid suspension. In this work, the minimum agitation speed, N_{js}, to achieve the off-bottom suspension of 150-µm glass particles was experimentally measured in a cylindrical tank stirred by a disk turbine for the case in which the liquid height-to-tank diameter ratio, *H/T*, is less than 1. Experiments were conducted with different impeller off-bottom clearances (C_b/T), impeller diameter ratios (D/T), and impeller submergence ratios (S_b/D). Off-bottom solid suspension was achieved at lower agitation speed when the impeller is closer to the tank bottom. More importantly, for each impeller off-bottom clearances there is a minimum liquid level below which the solids cannot be suspended (N_{js}®¥). When the impeller submergence ratio reaches a critical low value, solids can no longer be suspended even if the impeller speed is increased to a very high number. The critical S_b/D ratio for C_b/T=0.30 and 0.14 was 0.77 which is in agreement with the value of submergence ratio at which to the flow pattern transitions from either a Double-Loop regime to Single Loop-up Regime for C_b/T>0.16 or from the Single Loop-down Regime to Single Loop-up Regime for C_b/T<0.16, as previously found by our group. The results obtained here for N_{js} confirm that a stirred vessel operating at low impeller submergence is more likely to experience impeller flooding conditions caused by surface air entrainment, especially at high agitation speed. Under these conditions, the impeller loses its ability to provide adequate liquid pumping and recirculaton, and this results in poor mixing conditions especially near the vessel bottom, thus reducing the solid suspension capability of the impeller.

In Cooperation with Technische Universität Berlin,
Chair of Chemical and Process Engineering

SOPAT Smart Online Particle Analysis Technology

Sizing bubbles in turbulent systems

S. Maaß[1], R. Panckow[1,2], S. Ulrich[1], M. Kraume[2]

[1]SOPAT GmbH, Berlin, Germany; [2]Technische Universität Berlin, Chair of Chemical and Process Engineering, Berlin, Germany

Introduction

SOPAT Smart Online Particle Analysis Technology

- Significant focus on the ability to control fluid particle systems

Fermenter

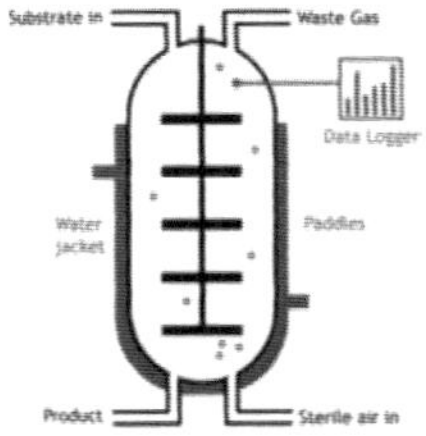

[1]

FPA reactor

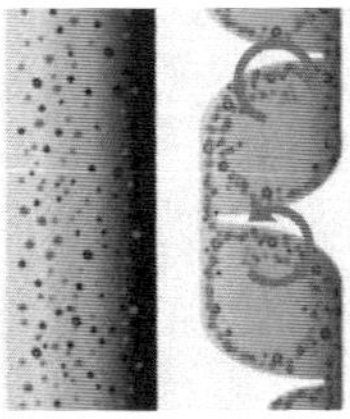

[2]

Flotation

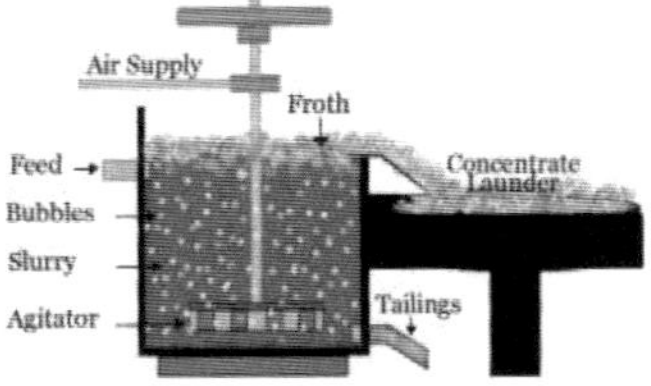

[3]

- Heat & mass transport optimization
- Particle boundaries (sizes & shapes) of major interest

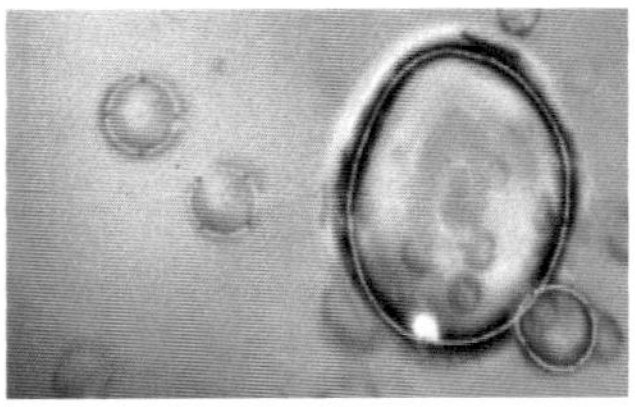

[1] http://scienceaid.co.uk/biology/micro/fermentation.html
[2] http://subitec.com/de/fpa-reaktor
[3] http://stonecrushermachine.org/Africa

Measurement comparison

SOPAT Smart Online Particle Analysis Technology

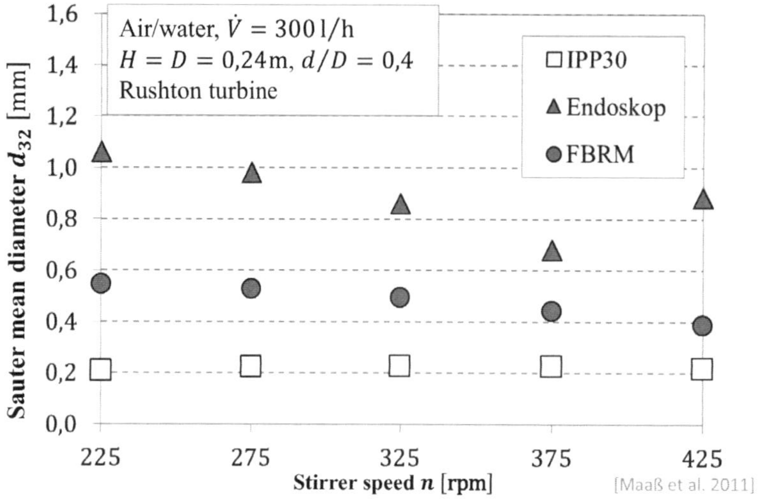

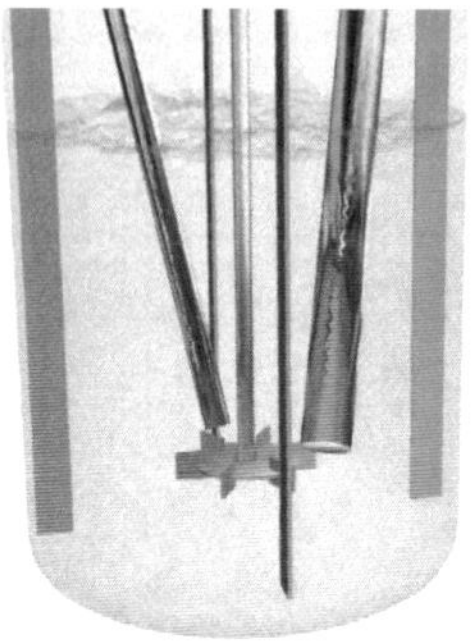

- Quantitativly and qualitativly hugh differences in the results
- Imaging based methods have the highest reliability

[Hu et al. 2006, Pacek et al. 1994]

Measurement comparison

SOPAT Smart Online Particle Analysis Technology

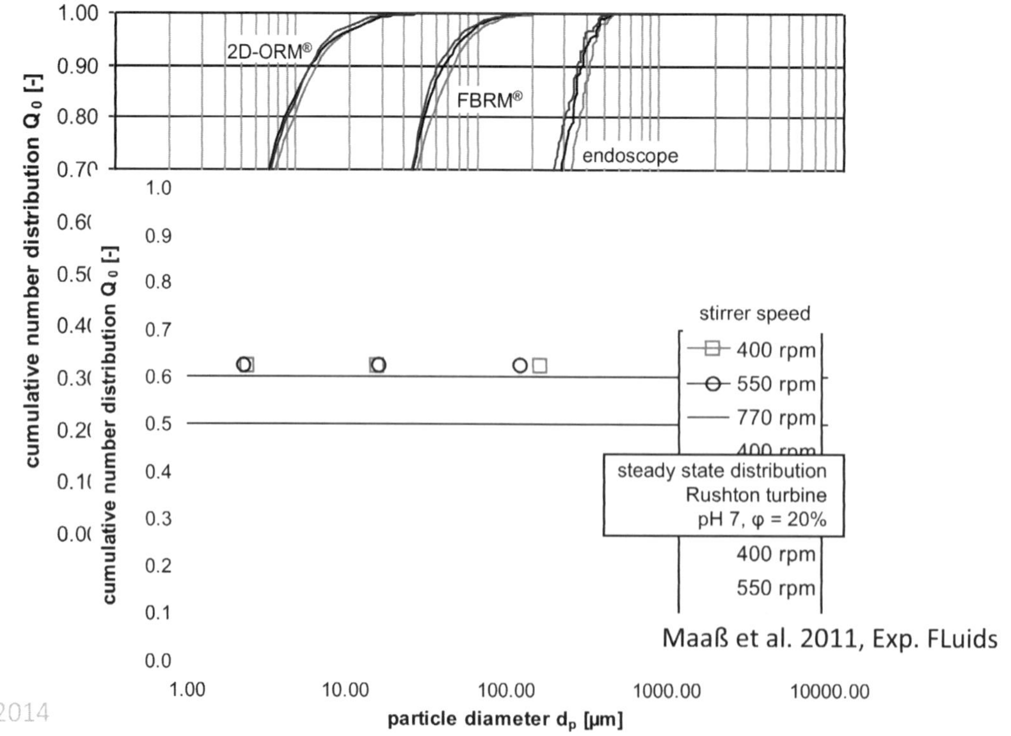

Maaß et al. 2011, Exp. FLuids

Statistical analysis

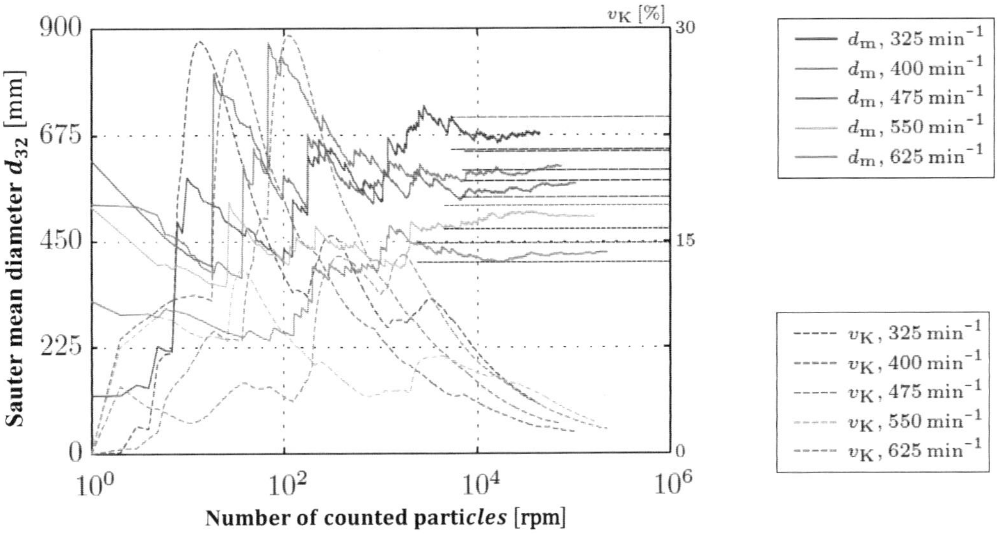

Experimental method

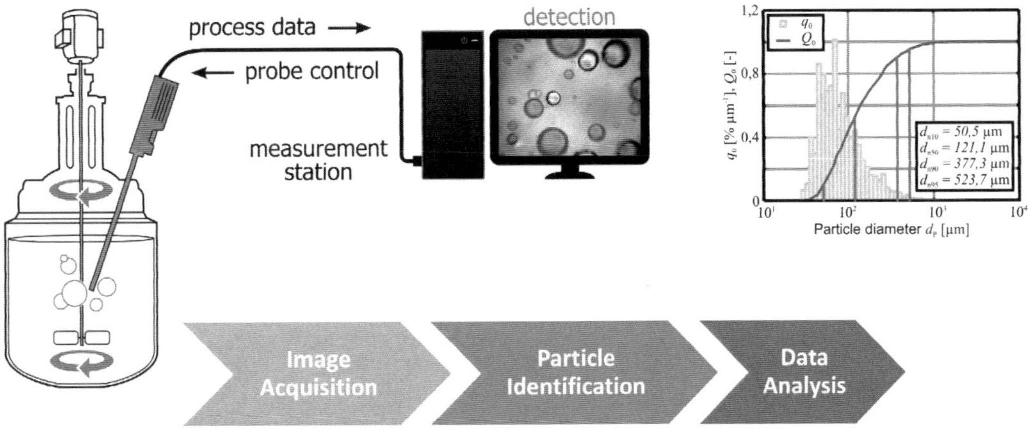

Working principle

SOPAT
Smart Online Particle Analysis Technology

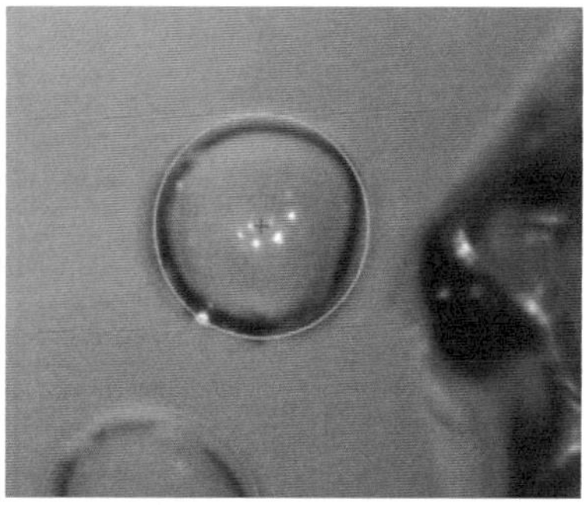

(1) mark some sample particles

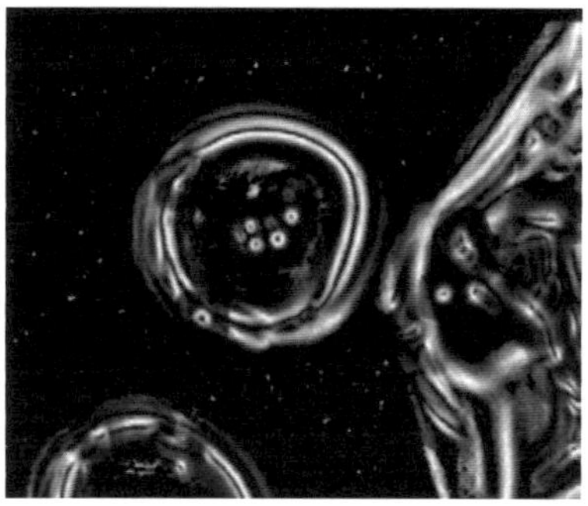

(2) Semi-automated generation of search pattern

(3) automatic image processing steps

Working principle

SOPAT
Smart Online Particle Analysis Technology

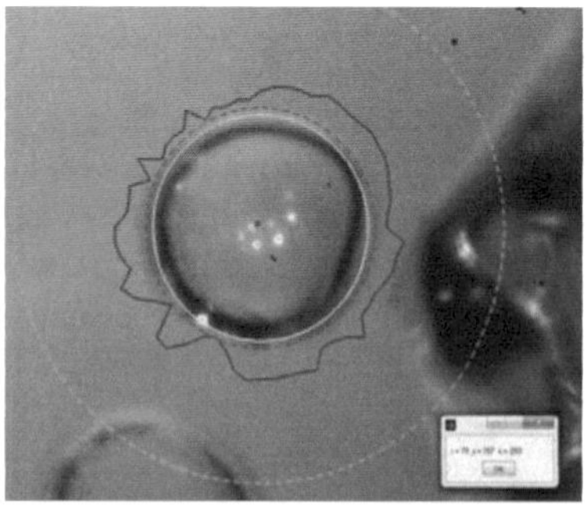

(4) particle candidates are classified

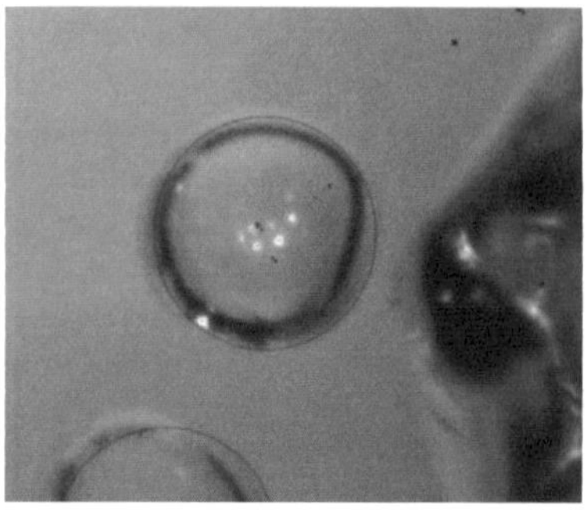

(5) particles are marked

(6) Generation of results

Detection of the software

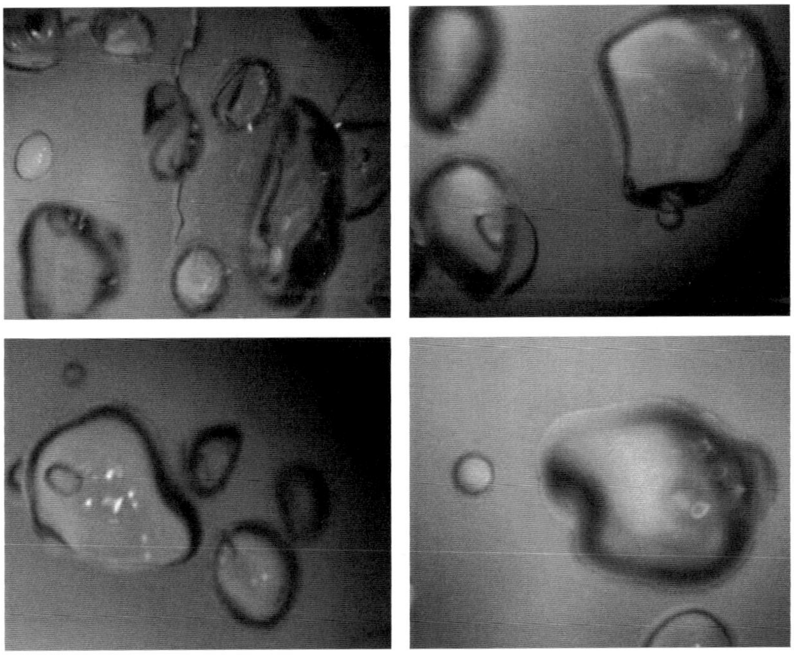

Experimental set-up

SOPAT Smart Online Particle Analysis Technology

- Comparison of two mixers
- Comparison of two differen volume streams

PHILADELPHIA MIXING SOLUTIONS, Ltd.

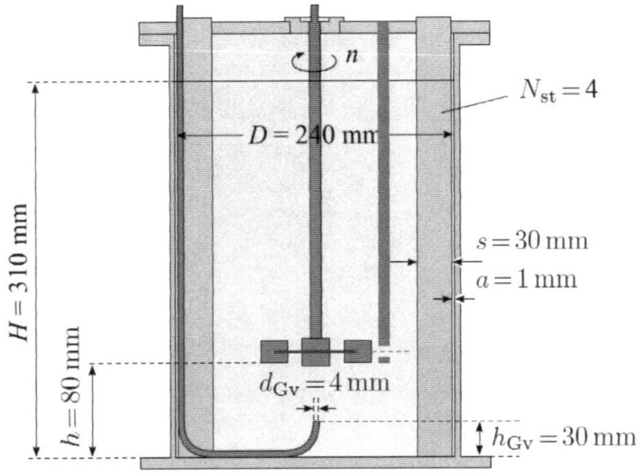

Results

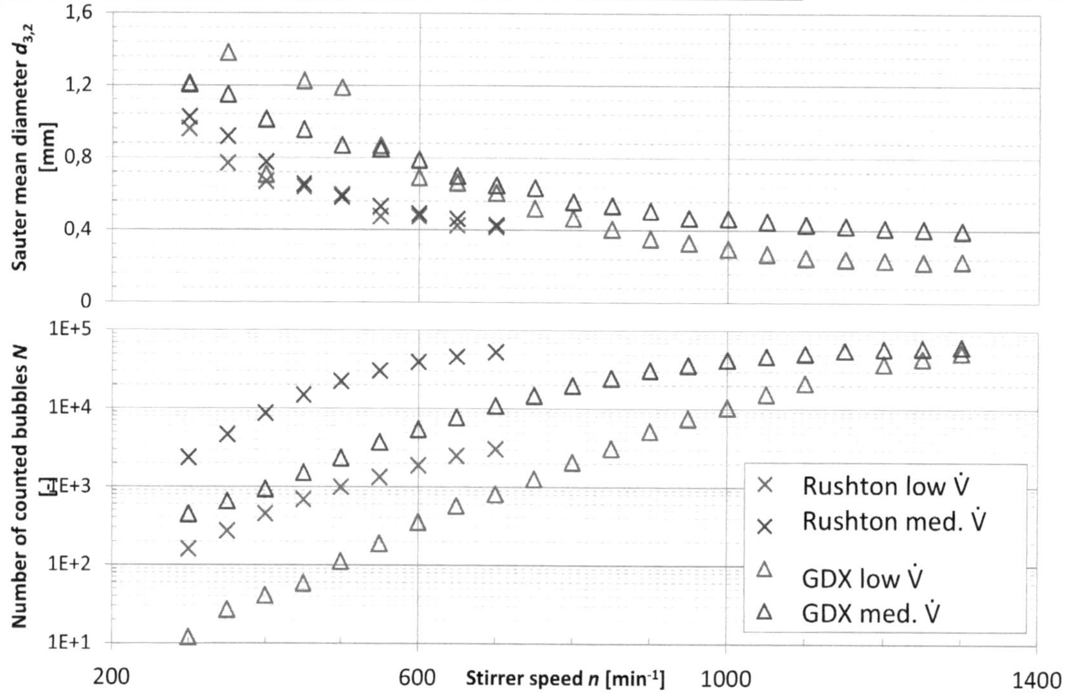

Results

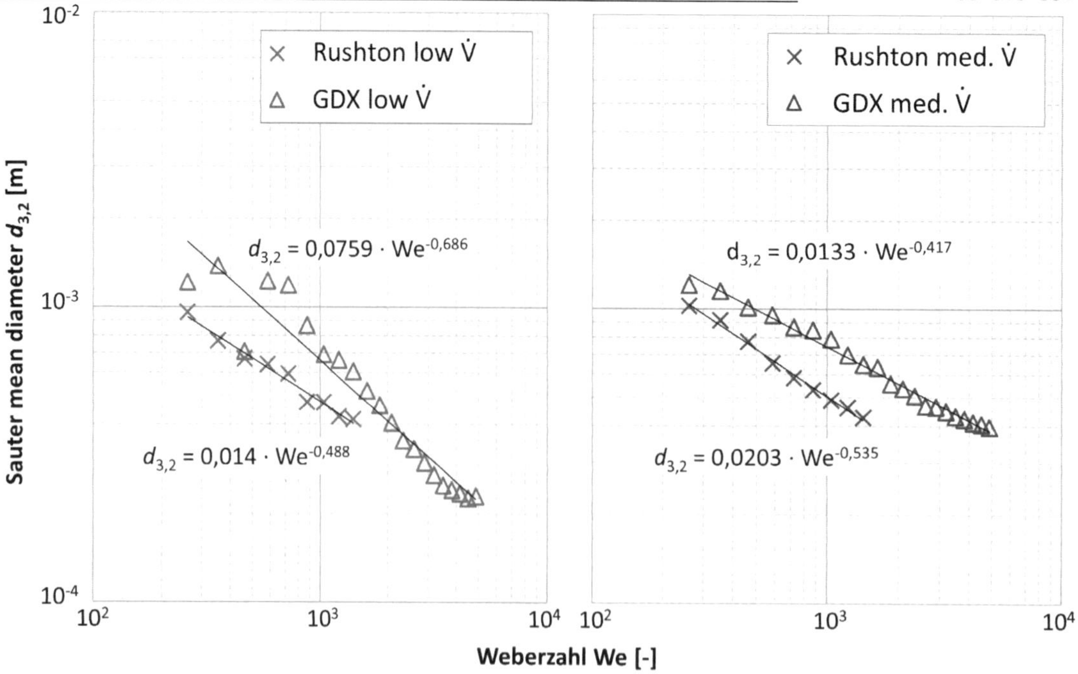

Outlook

SOPAT
Smart Online Particle Analysis Technology

- We measure happy bubbles
- We hardly work on realtime measurements decreasing the measurement time from minutes to seconds
- We are adapting the system to three and four phase systems
- Simulate these systems with CFD and PBE

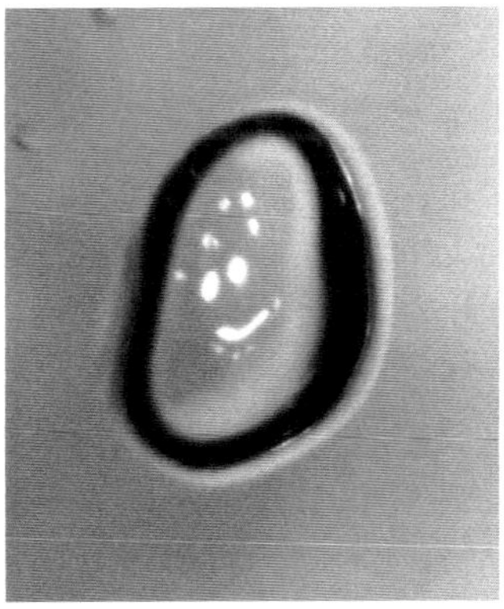

SOPAT
Smart Online Particle Analysis Technology

Interested?...

...Visit our ***booth 110***

P 38 – Results and discussion

SOPAT Smart Online Particle Analysis Technology

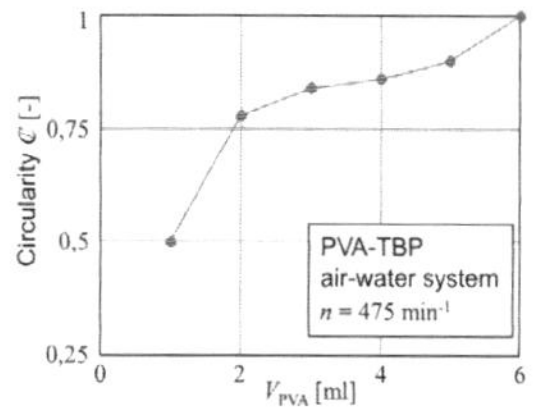

- Gas-liquid system (complex curved particles)
- 3 systems investigated (surface-active auxiliaries)
- Spherical shape is deviated from (↑ irregularity)

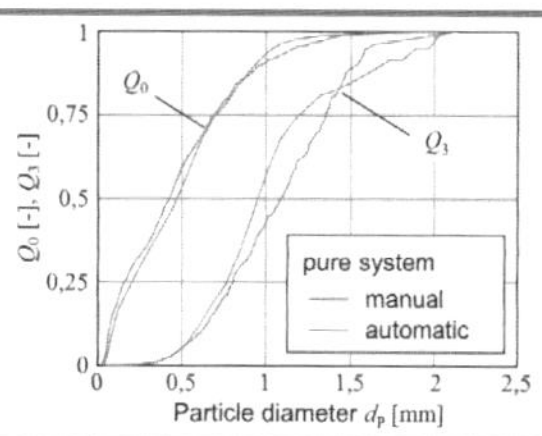

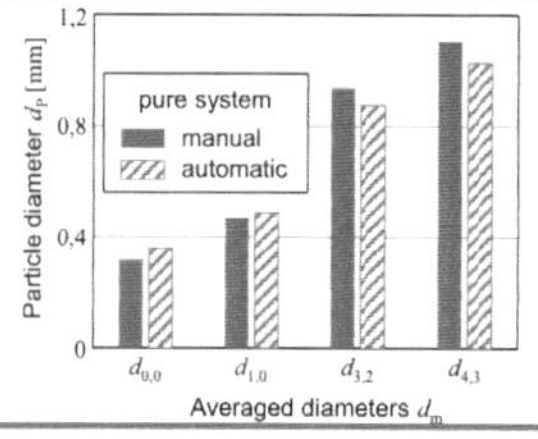

- Validating the results
- Comparison with manual counting

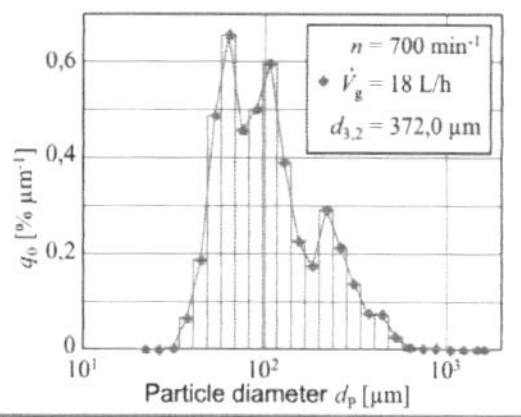

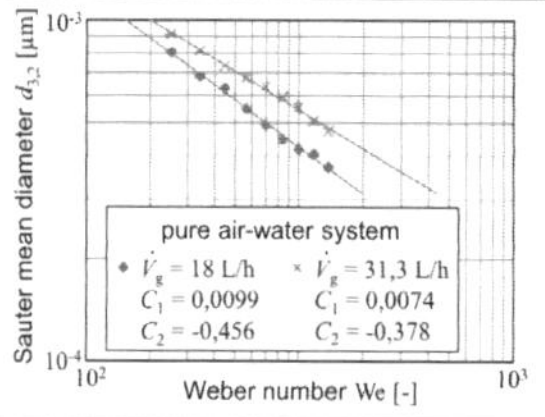

- Determine stable PSD
- Correlating the results (operating parameters)

mirror effect on clear toluene drops

SOPAT Smart Online Particle Analysis Technology

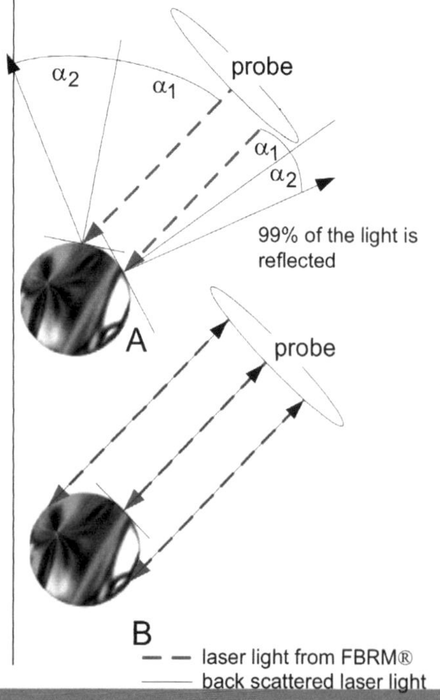

- ❖ drop sizes measured with the laser techniques are mostly too small
- ❖ back scattering of the laser beams are punctuated – mirror effect (A)
- ❖ only three points of the surface reflect the laser light back to the probe (B)

359798 Studying Mixing in Non-Newtonian Blue Maize Flour Suspension Using Color Analysis

Abstract:

Background: Non-Newtonian suspensions fluids occur in many relevant flow and mixing scenarios at the lab and industrial scale. The addition of acid or basic solutions to a non-Newtonian fluid is not an infrequent operation, particularly in Biotechnology applications where the pH of Non-Newtonian culture broths is usually regulated using this strategy.

Methodology and findings: We conducted mixing experiments in agitated vessels using Non-Newtonian blue maize flour suspensions. Acid or basic pulses were injected to reveal mixing patterns and flow structures and to follow their time evolution. No foreign pH indicator was used as blue maize flours naturally contain anthocyanins that act as a native, wide spectrum, pH indicator. We describe a novel method to quantitate mixedness and mixing evolution through Color Analysis (CA) in this system. Color readings corresponding to different times and locations within the mixing vessel were taken with a digital camera (or a colorimeter) and translated to the CIEL-*Lab* scale of colors. We use distances in the *Lab* space, a 3D color space, between a particular mixing state and the final mixing point to characterize segregation/mixing in the system.

Conclusion and relevance: Blue maize suspensions represent an adequate and flexible model to study mixing (and fluid mechanics in general) in Non-Newtonian suspensions using acid/base tracer injections. Simple strategies based on the evaluation of color distances in the CIEL-*Lab* space can be adapted to characterize mixedness and mixing evolution in experiments using blue maize suspensions.

377539 Experimental Determination of the Mixing Requirements for Solid Suspension in Pharmaceutical Stirred Tank Reactors

Abstract:

Glass and glass-lined, stirred-tank reactors are of significant importance in the pharmaceutical and related industries. Because of fabrication issues, a retreat blade impeller(RBI) with a low impeller clearance off the tank bottom is commonly used in glass-lined reactors, typically combined with a single baffle (providing only partial baffling conditions) mounted from the top of the reactor. In addition, these reactors are often provided with a torispherical bottom. Other configurations are also used, including full baffling or no baffles at all, hemispherical bottoms, and different impeller types. Despite their common use, some of the most important mixing characteristics of this type of reactor have not been fully studied, such as the minimum impeller agitation speed, N_{js}, to just suspend finely divided solids. In this work, N_{js} was experimentally obtained for a number of different pharmaceutically-relevant agitation systems including vessels with different types of bottoms (torispherical or hemispherical), impeller types (RBI, disk turbine, 4-blade and 6-blade pitched-blade turbine), baffling conditions (fully baffled, partially baffled, and unbaffled tanks), impeller clearance off the tank bottom, and solid particle size. The power dissipated by the impeller was also measured. A novel experimental approach to determine N_{js} was developed here. This approach consists of experimentally measuring the size of the circular region of the tank bottom covered by the solids at increasing values of the agitation speed, N, plotting N vs. the size of this region (expressed as either the region's diameter D_S or its area A_S), and linearly regressing the data to obtain N_{js} as the limit of the N value for D_S or A_S going to zero. The N_{js} results obtained with the new approach were compared with those obtained using the traditional Zwietering's approach, visually requiring that the solids do not rest on the tank bottom for more than 1-2 seconds. Excellent agreement was found between the results obtained using the novel approach and those obtained using Zwietering's method. The novel method proposed here completely eliminates the observer's bias from the experimental determination of N_{js}. The results obtained here show that N_{js} is a strong function of most of the variables listed above, and especially the baffling type.

392625 Vertical Solids Distribution in Agitated Vessels-Effect of Mixed Density and Particle Size Solids on Cloud Height-Initial Results

Abstract:

A simple measure of vertical solids distribution in agitated tanks is the cloud height, the demarcation between a lower solids rich height and a region almost clear of solids. A number of researchers have measured this height as a function of agitator speed and physical properties for uniform particle properties. In this study we look at the effect of a mixture of particles of different densities and particle sizes. Initial results show that the individual cloud heights are surprisingly not adversely affected by blending two different solids. In fact in some cases the total cloud height increases on blending two solids.

354947 An Impeller-Draft Tube Agitation System for Gas-Liquid Mixing in a Stirred-Tank Reactor

Abstract:

Stirred-tank reactors (STR) are widely used in chemical processes where intimate contact between a gas and a liquid is required to achieve good mass transfer. The design of the STR significantly affects gas bubble formation and dispersion, the gas-liquid interfacial area density, and the mass transfer coefficient. These in turn influence the rate of conversion, selectivity, and yield of the reaction. This work relates to the design of an agitation system for gas-liquid contacting in STRs that is critical to applications requiring higher gas flow rates. It incorporates a draft tube with one or more down-pumping axial-flow impellers located within. The draft tube has internal vertical baffles to disrupt tangential swirl and facilitate axial flow. A radial-flow impeller is located outside and below the bottom opening of the draft tube. The reactant gas is introduced below the radial-flow impeller in the conventional manner through a ring or nozzle sparger. The axial-flow impellers inside the draft tube pump the liquid downward in the core region of the STR, then the radial-flow impeller disperses the gas into the liquid and directs the resulting two-phase mixture radially outward so that it rises in the annular region and recirculates back into the draft tube at the top of the STR.

The performance of the agitation system described above was studied experimentally using water sparged with air in a 0.45 m diameter STR. The agitator consisted of two 0.13 m axial-flow hydrofoil impellers plus one 0.23 m radial-flow Rushton disk turbine at the bottom. In this study, the STR was configured both with and without a 0.30 m diameter draft tube for comparison of their flow characteristics and mass transfer efficacy. A transient method was used to determine the mass transfer coefficient $k_{L}a$. The water was initially stripped free of oxygen; then after the air sparging began, the transient dissolved oxygen concentration was measured and used to calculate $k_{L}a$. In addition to the mass transfer rate, the specific power was also measured. The results show that the incorporation of the draft tube in the STR yielded a system that has improved performance compared with conventional gas-liquid STRs without the draft tube, offering up to 20% increase in mass transfer at the same gas flow rate and agitation speed, or reducing the specific power draw by as much as 30% for similar $k_{L}a$ values at the same agitation speed.

Gas-Liquid Interfacial Area Production in a Vertical Column

Richard L. Long, Jr. PhD, P.E.
Professor Emeritus, NMSU
November 17, 2014

Approaches Used

- **Approach 1**: mass balances with no source or exchange terms, total momentum balance, gas phase momentum balance with source term models for forces
- **Approach 2**: mass balances with no source or exchange terms, momentum balance for each fluid phase with exchange terms between the phases, interfacial energy balance with source term from the continuous phase mechanical energy balance

Outline

- Objective
- The Derived Model for Gas-Liquid Bubble Flow.
- Applying the Model to Fully Developed Turbulent Flow Bubble Columns.
- Contributions.
- Questions

Objective

•To derive a model for gas-liquid bubble flow in the bubbly flow regime.

•Apply the model to fully developed turbulent flow bubble columns using Prandtl's mixing length turbulent model and validate the model.

The General derived Model for Gas-Liquid Bubble Flow

The model was derived by making:

1- Mass balance on the gas phase.
2- Mass balance on the liquid phase.
3- Total momentum balance.
4- Momentum balance on the gas phase.

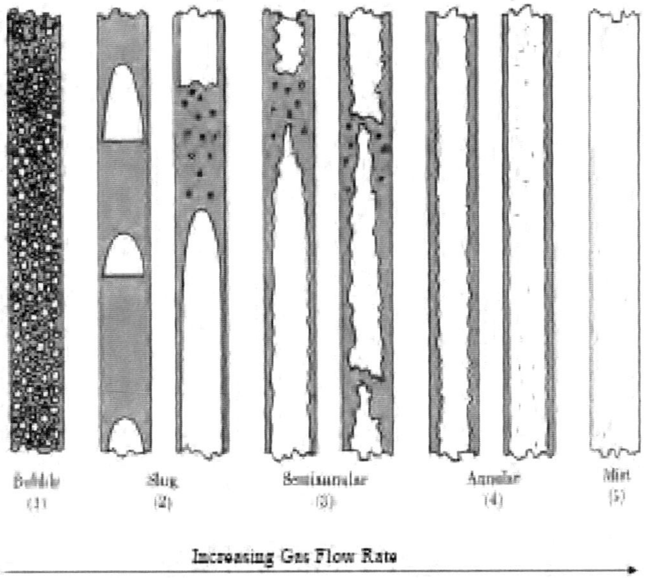

Figure 4.1: Flow Patterns in vertical Gas-Liquid Flow. [From Nevers, N. D., Fluid Mechanics for Chemical Engineers, Second Edition McGraw-Hill, (1991)].

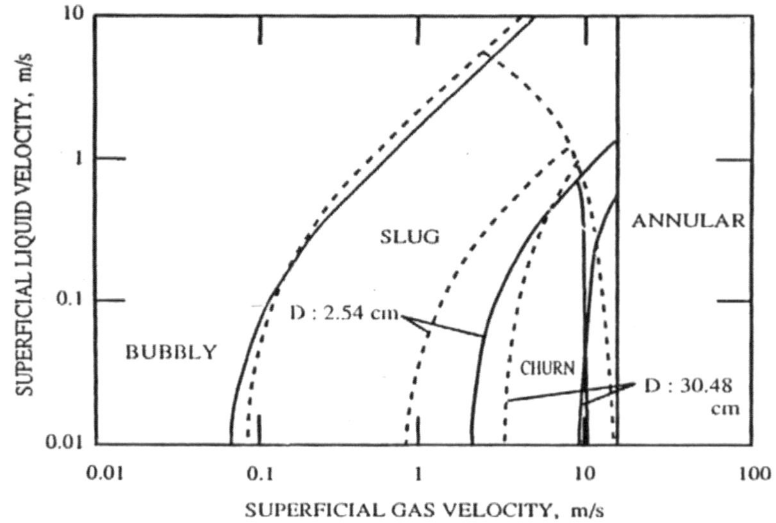

Flow region map for gas-liquid bubble flow in vertical tubes.

*****The hydrodynamic of each flow region in gas-liquid flow is completely different than the other regions*****

The main fact about gas-liquid bubble flow:

$$u_x^{gas} = u_x^{liquid}$$
$$u_y^{gas} = u_y^{liquid}$$
$$u_z^{gas} = u_z^{liquid} + V_{Bouyancy}^{gas}$$

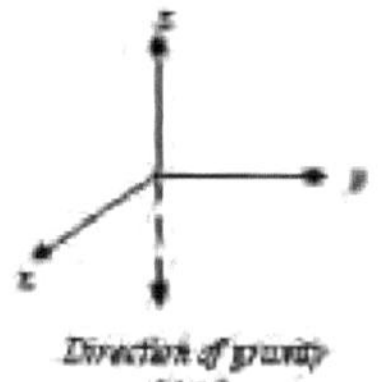

Or in vector form :

$$\vec{V}^{gas} = \vec{V}^{liquid} + V_{Bouyancy}^{gas}\,\vec{k}$$

Where,

$$V_{Bouyancy}^{gas} = \sqrt{\frac{4Dg}{3C_d}}$$

$\vec{k}$ is the unit vector opposite to the direction of gravity.

Mass Balance:

1) **Liquid phase:** $$\frac{\partial(\alpha_l \rho_l)}{\partial t} + \nabla.(\alpha_l \rho_l \vec{V}^{\text{liquid}}) = \dot{m}_{gl}$$

2) **Gas phase:** $$\frac{\partial(\alpha_g \rho_g)}{\partial t} + \nabla.(\alpha_g \rho_g \vec{V}^{gas}) = -\dot{m}_{gl}$$

Where :

α_l = void fraction of liquid phase

α_g = void fraction of gas phase

ρ_l = liquid density

$$\dot{m}_{gl} = \frac{\text{Rate of mass transfer from the gas phase to the liquid phase}}{\text{Unit volume}}$$

3) Total Momentum Balance:

$$\frac{\partial(\alpha_g \rho_g \vec{V}^{gas} + \alpha_l \rho_l \vec{V}^{liquid})}{\partial t} = -\nabla.(\alpha_g \rho_g \vec{V}^{gas}\vec{V}^{gas} + \alpha_l \rho_l \vec{V}^{liquid}\vec{V}^{liquid})$$
$$-\nabla.(\alpha_g \underline{\underline{\tau}}_l^{gas} + \alpha_l \underline{\underline{\tau}}_l^{liquid}) - \nabla.(\alpha_g \underline{\underline{\tau}}_t^{gas} + \alpha_l \underline{\underline{\tau}}_t^{liquid})$$
$$-\nabla P + (\alpha_g \rho_g + \alpha_l \rho_l)\vec{g}$$

Where

$$\underline{\underline{\tau}}_l = laminar\ stress\ tensor = -\mu(\nabla\vec{V} + [\nabla\vec{V}]^T) + \frac{2}{3}\mu\,\nabla.\vec{V}\,\underline{\underline{\delta}}$$

$$\underline{\underline{\tau}}_t = \text{turbulent stress tensor} = \rho\overline{\vec{V}'\vec{V}'}$$

Simplifying the mass and total momentum balance equations:

Because, the density and viscosity of liquids are much larger than the density and viscosity of gases:

$$\text{For gases}:\begin{cases}\rho_g \text{ is of order } 1\ kg/m^3\\ \mu_g \text{ is of order } 0.01\ cP\end{cases}$$

$$\text{For liquids}:\begin{cases}\rho_l \text{ is of order } 1000\ kg/m^3\\ \mu_l \text{ is of oredr } 1\ cP\end{cases}$$

For example at STP:

$$\rho_{air} = 1.2\ kg/m^3,\ \mu_{air} = 0.018\ cP \text{ and } \rho_{water} = 1000\ kg/m^3,\ \mu_{water} = 1\ cP$$

Therefore, the following four approximations can be made:

(1) $\alpha_l \rho_l + \alpha_g \rho_g \approx \alpha_l \rho_l$: For gas-liquid bubble flow usually $\alpha_g < 0.5 \Rightarrow \left(\dfrac{\alpha_g \rho_g}{\alpha_l \rho_l + \alpha_g \rho_g}\right)_{\alpha_g = 0.5} \times 100\% = 0.1\%$.

(2) $$\underbrace{\alpha_l \rho_l \vec{V}^{liquid}}_{\text{Liquid momentum accumulation per unit volume.}} + \underbrace{\alpha_g \rho_g \vec{V}^{gas}}_{\text{Gas momentum accumulation per unit volume.}} \approx \alpha_l \rho_l \vec{V}^{liquid}$$

(3) $$\underbrace{\nabla.(\alpha_l \rho_l \vec{V}^{liquid}\vec{V}^{liquid})}_{\text{Convection momentum of liquid phase per unit volume.}} + \underbrace{\nabla.(\alpha_g \rho_g \vec{V}^{gas}\vec{V}^{gas})}_{\text{Convection momentum of gas phase per unit volume.}} \approx \nabla.[\alpha_l \rho_l \vec{V}^{liquid}\vec{V}^{liquid}]$$

(4) $$\underbrace{\nabla.(\alpha_l \underline{\underline{\tau}}^{liquid})}_{\text{Rate of momentum gain by liquid viscous transfer per unit volume}} + \underbrace{\nabla.(\alpha_g \underline{\underline{\tau}}^{gas})}_{\text{Rate of momentum gain by gas viscous transfer per unit volume}} \approx \nabla.(\alpha_l \underline{\underline{\tau}}^{liquid})$$

Further, we assume:

(5) The density of the liquid phase is constant (ρ_l = Constant).

After substituting the five approximations in the mass and total momentum equations, we obtain:

Mass :-

$$Liquid: \quad \frac{\partial(\alpha_l)}{\partial t} + \nabla.(\alpha_l \vec{V}^{liquid}) = \frac{\dot{m}_{gl}}{\rho_l}$$

$$Gas: \quad \frac{\partial(\alpha_g \rho_g)}{\partial t} + \nabla.(\alpha_g \rho_g \vec{V}^{gas}) = -\dot{m}_{gl}$$

Total Momentum :-

$$\rho_l \frac{\partial(\alpha_l \vec{V}^{liquid})}{\partial t} = -\rho_l \nabla.(\alpha_l \vec{V}^{liquid} \vec{V}^{liquid}) - \nabla.(\alpha_l \underline{\underline{\tau}}_l^{liquid}) - \nabla.(\alpha_l \underline{\underline{\tau}}_t^{liquid}) - \nabla P + \alpha_l \rho_l \vec{g}$$

where,

$\underline{\underline{\tau}}_l^{liquid}$ = Liquid phase laminar stress-strain tensor

$$= -\mu_l \left[\nabla \vec{V}^{liquid} + (\nabla \vec{V}^{liquid})^T \right] + \frac{2}{3}\mu_l (\nabla.\vec{V}^{liquid}) \underline{\underline{\delta}}$$

$\underline{\underline{\tau}}_t^{liquid}$ = Liquid phase turbulent stress-strain tensor = $\rho_{liquid} \overline{\vec{V}'^{liquid} \vec{V}'^{liquid}}$

4) Gas phase momentum balance:

$$\frac{\partial(\alpha_g \rho_g \vec{V}^{gas})}{\partial t} = -\nabla.(\alpha_g \rho_g \vec{V}^{gas} \vec{V}^{gas}) + \vec{F}$$

Where, $\vec{F}$ = Sum of forces acting on the gas phase.

- The forces acting on the gas phase for laminar are:

1-Virtual mass force
2- Lift force
3- Turbulent Dispersion Force
4- Wall force
5-Drag force
6-Pressure force
7-Gravity force (negligible)

1 – *Virtual* mass [Drew and Lahey, 1987]:

$$\vec{F}_{vm} = C_{vm}\alpha_g \rho_l \left[\frac{D\vec{V}^{gas}}{Dt} - \frac{D\vec{V}^{liquid}}{Dt} \right]$$

2 – Lift Force [Drew and Lahey, 1987]:

$$\vec{F}_{lift} = \alpha_g C_l \rho_l (\vec{V}^{gas} - \vec{V}^{liquid}) \times \nabla \times \vec{V}^{liquid}$$

4 – *Turbulent* dispersion force [Carrica, Drew, Bonetto and Lahey 1999]:

$$\vec{F}_{td} = -\frac{3}{4}\rho_l \, v_t \frac{C_d}{D} \left| \vec{V}^{gas} - \vec{V}^{liquid} \right| \nabla\alpha_g$$

3 – The wall force [Antal, Lahey, and Flaherty, 1991]:

$$\vec{F}_{wall} = \frac{\alpha_g \rho_l}{D}\left[-0.208 + 0.147(\frac{D}{y})\right] \left| (\vec{V}^{gas} - \vec{V}^{liquid}) - [\vec{n}_w . (\vec{V}^{gas} - \vec{V}^{liquid})] \vec{n}_w \right|^2 \vec{n}_w$$

5 – Pressure force $= \alpha_g \nabla P_l$

6 – Drag force [Ishii and Mashima, 1984]: $\vec{F}_{drag} = -\frac{3}{4}\frac{\alpha_g}{D} C_d \rho_l (\vec{V}^{gas} - \vec{V}^{liquid}) \left| \vec{V}^{gas} - \vec{V}^{liquid} \right|$

7) Gravity force: The gravity force is: $\vec{F}_{gravity} = \alpha_g \rho_g \vec{g}$

4-Gas momentum For laminar flow:

After substituting the forces acting on the gas phase in the gas momentum equation, we obtain:

$$\frac{\partial(\alpha_g \rho_g \vec{V}^{gas})}{\partial t} = -\nabla.(\alpha_g \rho_g \vec{V}^{gas} \vec{V}^{gas})$$

$$-\frac{3}{4}\frac{\alpha_g}{D} C_d \rho_l (\vec{V}^{gas} - \vec{V}^{liquid}) \left| \vec{V}^{gas} - \vec{V}^{liquid} \right|$$

$$+\alpha_g C_{vm} \rho_l \left[\frac{D\vec{V}^{gas}}{Dt} + (\vec{V}^{gas} . \nabla)\vec{V}^{gas} - \frac{\partial \vec{V}^{liquid}}{\partial t} + (\vec{V}^{liquid} . \nabla)\vec{V}^{liquid} \right]$$

$$+\alpha_g C_l \rho_l (\vec{V}^{gas} - \vec{V}^{liquid}) \times \nabla \times \vec{V}^{liquid}$$

$$-\frac{3}{4}\rho_l \, v_t \frac{C_d}{D} \left| \overline{\vec{V}^{gas}} - \overline{\vec{V}^{liquid}} \right| \nabla\overline{\alpha_g}$$

$$+\frac{\alpha_g \rho_l}{D}\left[-0.208 + 0.147(\frac{D}{y})\right] \left| (\vec{V}^{gas} - \vec{V}^{liquid}) - [\vec{n}_w . (\vec{V}^{gas} - \vec{V}^{liquid})] \vec{n}_w \right|^2 \vec{n}_w$$

$$-\alpha_g \nabla P_l$$

Applying the Model to Fully Developed Turbulent Flow Bubble Columns

$$\begin{cases} \bar{\upsilon}_r^{gas} = \bar{\upsilon}_r^{liquid} = 0 \\ \bar{\upsilon}_\theta^{gas} = \bar{\upsilon}_\theta^{liquid} = 0 \end{cases} \Rightarrow \bar{\upsilon}_z^{liquid},\ \bar{\upsilon}_z^{gas} \ \textit{are functions of } r \textit{ only}$$

$$\textit{Mass Balance}: \begin{cases} \textit{Liquid}: & \frac{\partial}{\partial z}(\bar{\alpha}_{liq}\bar{\upsilon}_z^{liq}) = 0 \\ \textit{Gas}: & \frac{\partial}{\partial z}(\bar{\alpha}_{gas}\rho_{gas}\bar{\upsilon}_z^{gas}) = 0 \end{cases}$$

Total Momentum:

z-direction:

$$0 = \underbrace{-\frac{1}{r}\frac{\partial}{\partial r}[r\overline{\alpha_{liq}}(\tau_{rz})_{lam}^{liq}] - \frac{1}{r}\frac{\partial}{\partial r}[r\overline{\alpha_{liq}}(\tau_{rz})_{turb}^{liq}]}_{\textit{Shear force}} - \underbrace{\frac{\partial \bar{P}}{\partial z}}_{\textit{Pressure force}} - \underbrace{\overline{\alpha_{liq}}\rho_{liq} g}_{\textit{Gravity force}}$$

$$r\textit{-direction}: \quad \frac{\partial \bar{P}}{\partial r} = 0$$

$$\theta\textit{-direction}: \quad \frac{\partial \bar{P}}{\partial \theta} = 0$$

Gas Momentum:

z-direction:

$$0 = \underbrace{-\frac{3}{4}\frac{\bar{\alpha}_g}{D} C_d \rho_l \left(V_{Bouyancy}^{gas}\right)^2}_{\text{Drag force}} + \underbrace{\bar{\alpha}_g \frac{\Delta p_l}{\Delta z}}_{\text{Pressure force}}$$

r-direction:

$$0 = \underbrace{-\bar{\alpha}_{gas} C_l \rho_l \frac{\partial \bar{\upsilon}_z^{liq}}{\partial r}}_{\textit{Lift}\text{ force}} - \underbrace{\frac{\bar{\alpha}_g \rho_l V_{Bouyancy}^{gas}}{\bar{D}}\left[-0.208 + 0.147(\frac{\bar{D}}{y})\right]}_{\textit{wall}\text{ force}} - \underbrace{\frac{3}{4}\rho_l v_t \frac{C_d}{\bar{D}}\frac{\partial \overline{\alpha_g}}{\partial r}}_{\textit{Turbulent}\text{ dispersion force}}$$

θ-direction:

$$0 = \frac{\partial \bar{\alpha}_{gas}}{\partial \theta}$$

The pressure drop:

-Because $\frac{\partial \overline{P}}{\partial r}=0$ and $\frac{\partial \overline{P}}{\partial \theta}=0$. Therefore, $\overline{P}$ is independent of r and θ and the pressure drop is constant (i.e., $\frac{\partial \overline{P}}{\partial z}=\frac{\Delta \overline{P}}{\Delta z}$ = constant).

- From Bernoulli's equation:

$$-\frac{\Delta P}{\Delta z} = \text{Gravity head} + \text{Friction head} + \underbrace{\text{Dynamic head}}_{0} + \underbrace{\text{Work head}}_{0}$$

$$= \text{Gravity head} + \text{Friction head}$$

$$= \rho_m\, g + \frac{f}{d}\frac{\rho_m}{2}J^2$$

$$\Rightarrow \quad -\frac{\Delta P}{\Delta z} \approx \alpha_{liquid}^{average}\, \rho_l \left(g + \frac{f}{2d}J^2\right)$$

where,

$\rho_m = \alpha_{liquid}^{average}\rho_l + \alpha_{gas}^{average}\rho_g \approx \alpha_{liquid}^{average}\rho_l$, f = *friction factor* , J = *total superficial velocity*

Boundary Conditions and Constraints:

Boundary conditions:

1) $\left.\frac{d\overline{\upsilon}_z^{liquid}}{dr}\right|_{r=0} = \left.\frac{d\overline{\upsilon}_z^{gas}}{dr}\right|_{r=0} = 0$ (Symmetry at the center)

2) $\left.\overline{\upsilon}_z^{liquid}\right|_{r=R} = \left.\overline{\upsilon}_z^{gas}\right|_{r=R} = 0$ (No slip at the wall)

3) Because of the wall force, the bubbles cannot touch the wall and the wall is always wet by the liquid. Therefore,

$$\lim_{r\to R} \overline{\alpha}_{liquid} = 1 \quad or \quad \lim_{r\to R} \overline{\alpha}_{gas} = 0$$

Constraints:

1) $Q_l = \int_0^R \overline{\alpha}_{liquid}\, \overline{\upsilon}_z^{liquid}\, dA_c = 2\pi \int_0^R r\, \overline{\alpha}_{liquid}\, \overline{\upsilon}_z^{liquid}\, dr$

2) $Q_g = \int_0^R \overline{\alpha}_{gas}\, \overline{\upsilon}_z^{gas}\, dA_c = 2\pi \int_0^R r\, \overline{\alpha}_{gas}\, \overline{\upsilon}_z^{gas}\, dr$

Where, A_c = Column cross sectional area; Q_g and Q_l are the gas and the liquid volumetric flow rates, respectively.

Model Solution:

- The Prandtl's mixing length model for the Reynolds stress $(\tau_{rz})^{liq}_{turb}$ was used :

$$(\tau_{rz})^{liq}_{turb} = -\rho_{liq}\,\ell^2 \frac{d\bar{\upsilon}_z^{liq}}{dr}\left|\frac{d\bar{\upsilon}_z^{liq}}{dr}\right|$$

$$\nu_t = \ell^2 \left|\frac{d\bar{\upsilon}_z^{liq}}{dr}\right|$$

Where, ℓ is the mixing length which is function of position.

- The following model for the mixing length ℓ was used [Schlichting 1968]:

$$\ell = R\ [0.14 - 0.08(\tfrac{r}{R})^2 - 0.06(\tfrac{r}{R})^4]$$

Model Validation

- Validation of the model was accomplished by comparison its prediction against experimental data by:

[T. J. Liu and S. G. Bankoff, "Structure of Air-Water Bubbly Flow in a Vertical Pipe-I. Liquid Mean Velocity and Turbulence Measurements", Int. J. Heat and Mass Transfer, 36 (4):1049-1060:1993 and T. J. Liu and S. G. Bankoff, "Structure of Air-Water Bubbly Flow in a Vertical Pipe-II. Void fraction, Bubble Velocity and Bubble Size Distribution", Int. J. Heat and Mass Transfer, 36 (4):1061-1072:1993]

[Leung,W., Revankar, S.T., Y. Ishii, and M. Ishii"Axial Development of Interfacial Area and void concentrations profiles measured by double-sensor method"Int j Heat and Mass Transfer,38,3, pp. 445-453(1995).]

Note: Fortran was used to solve the model equations with $\Delta(r/R) = 0.01$

Final Model Equations for Fully developed Turbulent Flow Bubble Columns

1) $$\frac{d\bar{\upsilon}_z^{liq}}{dr} = \frac{v_l - \sqrt{v_l^2 + \ell^2 \frac{\alpha_{liq}^{average}}{\bar{\alpha}_{liq}} \frac{f}{d} J^2\, r}}{2\ell^2}$$

2) $$\frac{d\bar{\alpha}_{gas}}{dr} = \frac{4}{3}\frac{C_l}{C_d}\bar{\alpha}_{gas}\frac{\bar{D}}{\ell^2} + \frac{8\bar{\alpha}_{gas}}{3C_d}\frac{V_{Bouyancy}^{gas}\left[0.208 - 0.147(\frac{\bar{D}}{y})\right]}{v_l - \sqrt{v_l^2 + \ell^2 \frac{\alpha_{liq}^{average}}{1-\bar{\alpha}_{gas}} \frac{f}{d} J^2\, r}}$$

3) $$\bar{a}_i\,(r) = \frac{8g}{C_d\left(V_{Bouyancy}^{gas}\right)^2}\bar{\alpha}_{gas}(r)$$

4) $$V_{Bouyancy}^{gas} = \frac{\left(J - \frac{2}{R^2}\int_0^R r\bar{\upsilon}_z^{liquid}\,dr\right)}{\alpha_{liq}^{average}}$$

5) $$\bar{\upsilon}_z^{gas}(r) = \bar{\upsilon}_z^{liquid}(r) + V_{Bouyancy}^{gas}$$

6) $$\bar{\alpha}_{gas}(r) + \bar{\alpha}_{liquid}(r) = 1$$

7) $$-\frac{\Delta P}{\Delta z} \approx \alpha_{liquid}^{average}\,\rho_l\,(g + \frac{f}{2d}J^2)$$

where,

$$\frac{1}{\sqrt{f}} = 2.0\log_{10}(\mathrm{Re}_{pipe}^{mixture}\sqrt{f}) - 0.8$$

$$C_d = \frac{24}{\mathrm{Re}_{bubble}}\left(1 + 0.1\mathrm{Re}_{bubble}^{0.75}\right)$$

$$\mathrm{Re}_{bubble} = \alpha_g \frac{\rho_l V_{Bouyancy}^{gas} D}{\mu_l}$$

$$\mathrm{Re}_{pipe}^{mixture} = \frac{\rho_m J d}{\mu_m} \approx \frac{\rho_l J d}{\mu_l}$$

Boundary conditions:

$$\bar{\upsilon}_z^{liquid}(r = 0) = 0 \quad and \quad \lim_{r\to R}\bar{\alpha}_{gas} = 0$$

Unknowns:

$\bar{\upsilon}_z^{liq}(r)$, $\bar{\upsilon}_z^{gas}(r)$, $\bar{\alpha}_{gas}(r)$, $\bar{\alpha}_{liq}(r)$, $\bar{a}_i(r)$, $V_{Bouyancy}^{gas}$, and $\frac{\Delta P}{\Delta z}$

Results

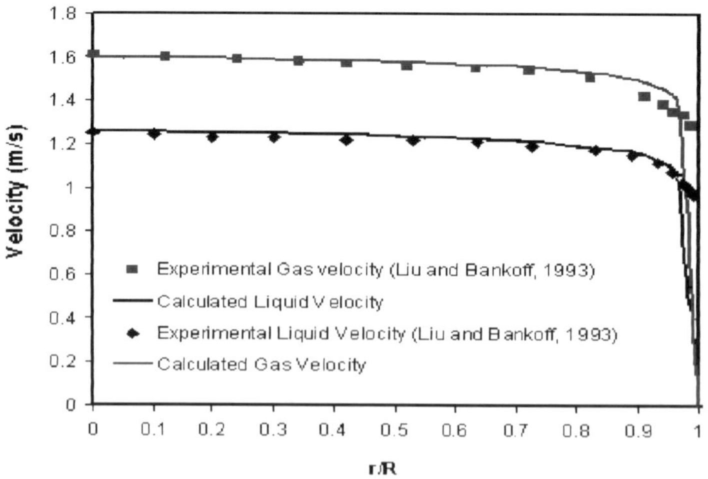

A comparison between the experimental values and the Calculated values of the radial profile of the gas velocity and the liquid velocity for air-water bubble flow at J_l =1.087 m/s and J_g =0.067 m/s.

Approach 1 Results for Vertical Bubble Column

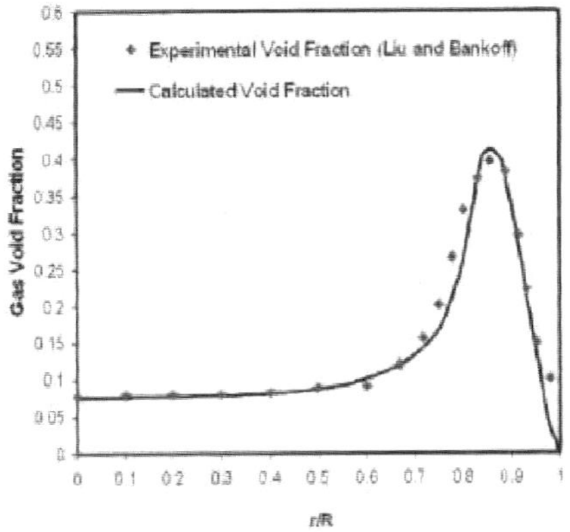

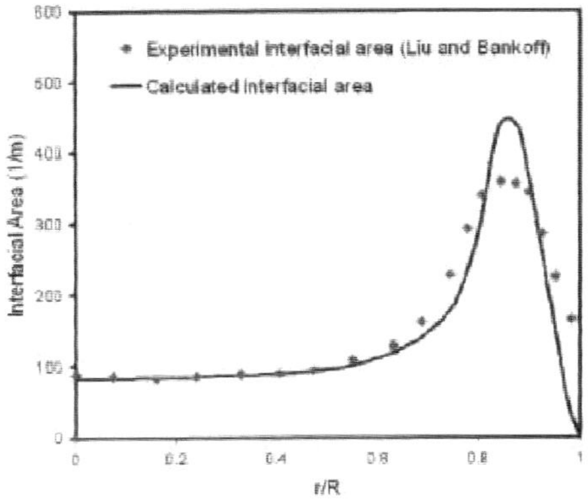

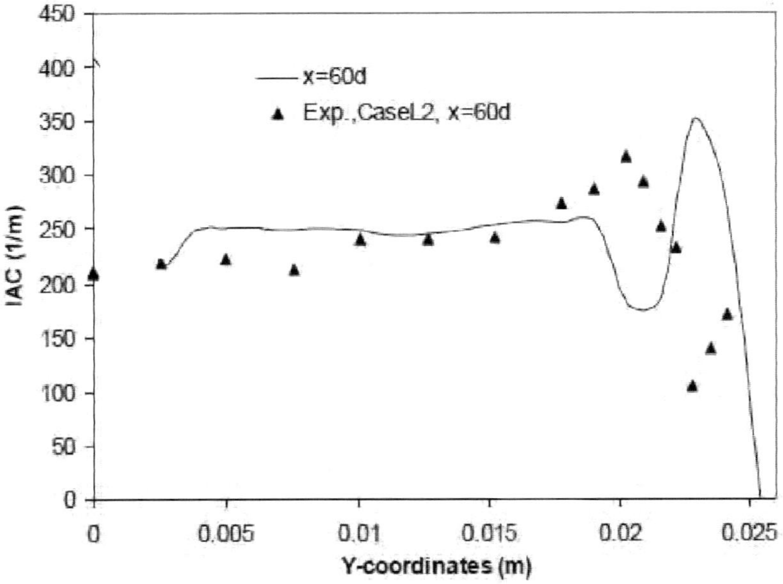

Contributions

1) A general model for gas-liquid bubble flow was derived.

2) A total momentum balance equation was derived instead of liquid momentum balance which simplifies the model.

3) A constitutive relation for the pressure drop was developed.

4) Simple turbulent model (The Prandtl's mixing length model) was used.

Conclusions

- **Approach 1** works fairly well. It is limited to gas-liquid systems and it relies on a simple buoyancy model and it is restricted to simple geometries.
- **Approach 2** still has some issues with it, but it holds the prospect of a broader scope of application.

References

- S.L. Yarbro and R.L. Long, "Using a New Interfacial Area Transport Equation to Predict Interfacial Area in Co-current Jet Mixers", Can. J. of Chemical Engineering, v.80, pp.622-630, August,2002.
- S. L. Yarbro, "Modeling Interfacial Area Transport in Multi-Fluid Systems" PhD Dissertation, NMSU, December,1996.
- M.A. Aliedeh, "Computational Fluid Dynamics Simulation of Interfacial Area Concentration in a Bubble Column" PhD Dissertation, NMSU, August, 2004 (**Approach 2**)
- A. al Jarrah, "Modeling Gas-Liquid Bubble Flow and Computational Fluid Dynamics Simulation of Interfacial Area Concentration in Bubble Columns", PhD Dissertation, NMSU, December, 2006 (**Approach 1**)

Questions?

377518 Power Dissipation and Power Numbers for a Retreat-Blade Impeller in a Torispherical-Bottomed Pharmaceutical Reactor Vessel Under Different Baffling Configurations

Abstract:

In the pharmaceutical industry, glass-lined reactors and vessels are often utilized to carry out a variety of different unit operations and especially reactions. Within these systems, the vessel impeller, and baffle(s) are typically glass-lined in order to provide superior corrosion resistance, prevent product contamination, and enhance cleanability. Because of the fabrication issues associated with glass lining metal surfaces, including baffles and impellers, these reactors are typically equipped with a single retreat-blade impeller (RBI) placed close to the tank bottom and a single baffle mounted from the top. Despite their common use in industry, these systems have received relatively little attention.

In this study, the power dissipation and the Power Number, Po, were experimentally obtained and computationally predicted in a 60-L scale-down version of a full-scale reactor for pharmaceutical production provided with a torispherical bottom, a RBI, and a single beavertail baffle, in order to simulate a geometrically similar production reactor. In addition, different baffling configurations were studied, i.e., fully baffled and unbaffled systems. Water and sucrose solutions of different viscosity were used to examine a wide range of Reynolds Numbers (Re). At lower Re values (Re~100) all systems produced similar Po values irrespective of baffling. However, as Re increased, Po for the unbaffled and, to a more limited extent, unbaffled systems decreased appreciably with respect to the Po for the fully baffled system. Po was found to be constant at 0.73 for the fully baffled system at high Re. For the unbaffled system Po decreased with Re over the entire Re range explored here (100<Re<300,000), asymptotically approaching the 0.2-0.25 value. Po for the partially baffled system was intermediate between the other two cases, with Po approaching the ~0.5 value at high Re.

Computational Fluid Dynamics (CFD) was used to model all the above-mentioned systems, but only at high Re, using different modeling approaches including single reference frame (SRF), multiple reference frame (MRF), and sliding mesh (SM) models. The k-ε turbulence model was used in all cases. The CFD simulations resulted in the prediction of the power dissipated by the impeller and therefore Po. These predictions were compared with the experimental results. In the fully baffled system, the predicted Po values using MRF modeling were in close agreement with the experimental results. In the partially baffled system, the results obtained with MRF modeling were very consistent with the experimental results. However, even better agreement was obtained when the much more computationally expensive SM modeling was used. Finally, the simpler SRF approach proved to be very appropriate to model the unbaffled system, and good agreement between the simulation predictions and the experimental results was obtained, but only if the surface deformation of the liquid-air interface typically observed in unbaffled systems was small.

Finally, the Po-vs.-Re experimental data were additionally fitted with a semiempirical equation whose parameters were adjusted for each baffling system. The resulting equations closely matched the experimental data.

NOV Process and Flow Technologies
5870 Poe Avenue
Dayton, OH 45414
Phone: 937-454-3200
Fax: 937-454-3379

Effective Static Mixer Blending

In water and wastewater treatment, the efficient mixing of small amounts of additives into a main flow with low energy is very important. These additive materials may be pH adjustments, flocculants and other polymers, amongst others. Static mixers are often used to achieve this mixing.

Blending two fluid streams using static mixers is affected by a variety of parameters such as the flow rates, density ratio and viscosity ratio of the streams. The difficulty of the blending operation increases as each of these ratios increases. Further, design of a static mixer for effective blending is often dictated by balancing a number of factors including permissible pressure drop, the maximum downstream distance over which blending must be achieved and the physical length of the mixer. Historically, mixers with low mixing distances and high pressure drops or mixers with long mixing distances and low pressure drops have been available.

This presentation will trace the development, characterization, and application of the UltraTab static mixer. The optimization of the injectant flow parameters will also be discussed. For turbulent flow, this mixer was developed to provide a coefficient of variation in composition of 0.05 within three pipe diameters downstream of the mixer even when blending streams with significantly different flow rates, viscosities, and densities.

Numerical Simulation of Micro-mixing in Stirred Reactors Using the Engulfment Model Coupled with CFD

Xiaoxia Duan, Xin Feng, Chao Yang, Zai-Sha Mao
(Institute of Process Engineering, Chinese Academy of Sciences, Beijing 100190, China)

Abstract

Stirred tanks are widely encountered in process industries, often used for complex and fast reactions such as reactive crystallization, nitrification and sulfonation. The yield and selectivity of desired products depend directly on the conditions of mixing on the molecular level. The micro-mixing simulations are being developed with improvement of computer capacity and application of computational fluid dynamics. The engulfment model (E-model) based on the turbulence promoted diffusion has been widely used. Owing to the computational efficiency and model simplicity, the E-model compared with other micro-mixing models based on the probability density function is more suitable for industrial scale reactors. However, the E-model does not contain the information about macro-mixing on the reactor scale and is only resolved in the Lagrangian frame perspective. In this paper, a CFD method combining with the E-model is presented and implemented to predict the micro-mixing effects on the course of parallel competing chemical reactions carried out in a semi-batch single phase stirred tank. The new modeling approach describes the small-scale segregation of fluid by solving the transport equations of mixture fraction and its variance based on the CFD simulation of the macro-flow, and then calculates the representative turbulence kinetic energy and dissipation in the reaction zone via the weighted average of the mixture-fraction variance for the E-model. The model predictions of segregation index at different agitation speeds and feeding locations are compared with the experimental results in satisfactory agreement.

Keywords: stirred tank; engulfment model; micro-mixing; numerical simulation; CFD;

1. Introduction

Stirred tanks are widely used in process industries, such as petroleum, chemical processing, pharmacy and metallurgy. The interaction between the stationary baffles and the rotating impellers produces complicated turbulent flow. The flow characteristics affect the mixing process. Mixing mechanisms can be divided into three main categories (i.e., macro-mixing, meso-mixing, micro-mixing) according to length scale. Chemical reaction is a molecular-level process, so the molecular-scale micro-mixing can affect it directly. Process industries usually involve complex fast reactions, and the yields of desired products and the distributions of products not only depend on chemical kinetics, but also the mixing especially micro-mixing when reaction time is shorter than or the same order of the time required bringing homogeneity at molecular scale. Poor micro-mixing will produce more unwanted products leading to higher purification costs. The investigations of micro-mixing contribute to the reasonable regulation of reactions and reduction of side products[1].

Unfortunately, the micro-mixing on the molecular scale cannot be resolved straightway due to the limitation of grid size in CFD simulation. Thus, the flow field is firstly calculated with CFD, and then the simulations of micro-mixing and chemical reactions are carried out with the probability density function methods (Finite-mode PDF[2], Multiple-time-scale turbulent mixer model[3], Direct quadrature method of moments combining with the interaction by exchange with the mean micro-mixing model[4]) or the Engulfment model[5,6]. The E-model is based on the turbulence theory[7] and widely used owing to the computational efficiency and model simplicity. The model can be used to predict micro-mixing efficiency when the turbulent energy dissipation rate in the reaction zone is known. Whereas the turbulent energy dissipation rate of the feeding point, rather than that of reaction zone, is used to calculate the E-model, this approach does not contain the effect of macro-mixing on the complex chemical reactions in the reactor.

In this work, a new micro-mixng simulation method related the E-model with CFD model is described in detail and is applied to investigate the micro-mixing in the single-phase stirred tank. The aim is to predict the micro-mixing effects on the course of parallel competing reactions carried out in different conditions. The numerical method is validated through the comparison between the predicted results and the reported experimental data.

2. E-model coupling to CFD

The parallel competitive reactions with second-order kinetics are considered as follows:

$$\mathrm{A} + \mathrm{B} \xrightarrow{k_1} \mathrm{R} \tag{2-1}$$

$$\mathrm{A} + \mathrm{C} \xrightarrow{k_2} \mathrm{S}$$

In most chemical engineering applications, the first reaction is an instantaneous reaction and the second reaction is a finite-rate reaction. We can introduce the mixture fraction and reaction-progress variables to describe the reaction system[8]. The mixture fraction, a non-reacting scalar, is related to the local concentrations of reactants A, B and C defined by

$$\xi = \frac{c_{\mathrm{A}} - c_{\mathrm{B}} - c_{\mathrm{C}} + c_{\mathrm{B0}} + c_{\mathrm{C0}}}{c_{\mathrm{A0}} + c_{\mathrm{B0}} + c_{\mathrm{C0}}} \tag{2-2}$$

where c_{A0}, c_{B0} and c_{C0} are the initial concentrations of reactants A, B and C. The mixture fraction, ranging from 0 to 1, reflects the turbulent mixing degree of different reactants[9].

In the CFD simulation, we can get ξ at every spatial location in the reactor only by solving its microscopic transport equation with use of direct numerical simulation (DNS). However the computational cost is enormous, this model is limited to predict the turbulent flow in the simple geometry reactor under low Reynolds numbers. Instead, the Reynolds-averaged Navier-Stokes (RANS) model can be used to describe the turbulent flow and yield the ensemble-averaged mixture fraction $\langle \xi \rangle$, which describes macro-mixing. Moreover, the degree of local mixing needs another variable, *i.e.*, the mixture-fraction variance $\langle \xi'^2 \rangle$, to measure in the case of the RANS model[9]. The mixture-fraction variance $\langle \xi'^2 \rangle$ is defined by

$$\langle \xi'^2 \rangle = \overline{(\xi - \langle \xi \rangle)^2} \tag{2-3}$$

The mixture-fraction variance is a measure for the small-scale segregation of the fluid, which represents the level of turbulent mixing[10]. In the perfectly developed turbulent mixing region, no segregation exists and the value of $\langle \xi'^2 \rangle$ is zero. For the semibatch mode, the limiting reactant is slowly added into the reactor and the segregation concentrates on a small zone close to the feed point[11]. If the reaction is an instantaneous reaction (or fast reaction), it is often localized to the zone near the feed pipe. Thus, through solving the transport equations of $\langle \xi \rangle$ and $\langle \xi'^2 \rangle$, we can get the distribution of the mixture-fraction variance, which quantifies the segregation of materials and provides information on the turbulent mixing in the reactor.

The Reynolds-average transport equations for the mean mixture fraction $\langle \xi \rangle$ and mixture-fraction variance $\langle \xi'^2 \rangle$ are given by[12]

$$\frac{\partial \langle \xi \rangle}{\partial t} + \langle u_j \rangle \frac{\partial \langle \xi \rangle}{\partial x_j} - \frac{\partial}{\partial x_j}\left(\Gamma_t \frac{\partial \langle \xi \rangle}{\partial x_j} \right) = 0 \tag{2-4}$$

$$\frac{\partial \langle \xi'^2 \rangle}{\partial t} + \langle u_j \rangle \frac{\partial \langle \xi'^2 \rangle}{\partial x_j} - \frac{\partial}{\partial x_j}\left(\Gamma_t \frac{\partial \langle \xi'^2 \rangle}{\partial x_j} \right) = 2\Gamma_t \left| \nabla \langle \xi \rangle \right|^2 - 2\gamma \langle \xi'^2 \rangle \tag{2-5}$$

where Γ_t is the turbulent diffusivity coefficient and γ is the micro-mixing rate. The detailed description can be referred to the literature[13].

The turbulence kinetic energy and dissipation in each cell can be predicted by the CFD simulation of the macro-flow. According to Eq.(2-6), we can calculate the average values of turbulence kinetic energy ($\bar{k}$) and dissipation ($\bar{\varepsilon}$) in the reaction zone via the mixture-fraction variance weighted average in the whole computational domain for the E-model.

$$\bar{k} = \frac{\sum \langle \xi'^2 \rangle_{\text{cell}} \langle k \rangle_{\text{cell}}}{\sum \langle \xi'^2 \rangle_{\text{cell}}}, \quad \bar{\varepsilon} = \frac{\sum \langle \xi'^2 \rangle_{\text{cell}} \langle \varepsilon \rangle_{\text{cell}}}{\sum \langle \xi'^2 \rangle_{\text{cell}}} \tag{2-6}$$

where $\langle k \rangle_{\text{cell}}$ and $\langle \varepsilon \rangle_{\text{cell}}$ are the values of turbulence kinetic energy and dissipation in each cell, respectively; $\langle \xi'^2 \rangle_{\text{cell}}$ is the value of the mixture-fraction variance in each cell. For the parallel competing reactions used to investigate the micro-mixing, the side reaction occurs mainly in the large segregation region.

3. Numerical simulation

3.1 *Single-phase stirred tank setup and reaction system*

The stirred tank studied in this work is the same as that of Akiti et al.[5]. The tank (inner diameter *T*=290 mm) is a flat-bottom cylindrical vessel (Fig.1), with a liquid height of *H*=*T*. Four baffles (width *B*=*T*/10) are vertically fitted on the tank wall. The reactor is equipped with a downward-pumping, six-blade, 45° pitched blade turbine with diameter *D*=102 mm and the vertically projected height is 12 mm, located at *C*=*T*/3 off the tank bottom. There are three feeding positions, located at the middle plane between two adjacent baffles. The coordinates of the feed points are given in Table 1.

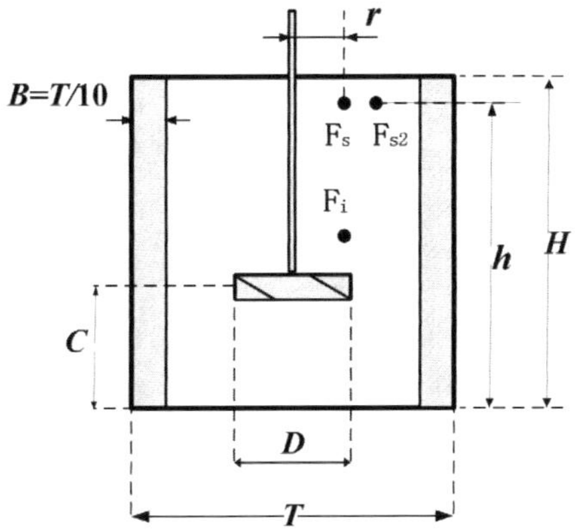

Fig.1 Details of the stirred tank

Table 1 Feeding positions

	F_s	F_{s2}	F_i
r/T	0.172	0.25	0.172
h/H	0.9	0.9	0.48

The reaction system employed in this work for characterizing mixing efficiency is the hydrolysis of ethyl chloroacetate in competition with the neutralization of sodium hydroxide:

$$\underset{(A)}{NaOH} + \underset{(B)}{HCl} \xrightarrow{k_1} \underset{(R)}{NaCl} + \underset{(P)}{H_2O} \tag{3-1}$$

$$\underset{(A)}{NaOH} + \underset{(C)}{CH_2ClOOC_2H_5} \xrightarrow{k_2} \underset{(Q)}{CH_2ClCOONa} + \underset{(S)}{C_2H_5OH} \tag{3-2}$$

The reaction rate constant k_1 is equal to 1.3×10^8 $m^3\cdot mol^{-1}\cdot s^{-1}$ and k_2 is 0.0257 $m^3\cdot mol^{-1}\cdot s^{-1}$ at 23 °C. The stirred tank contained initially the premixture of HCl (B) and $CH_2ClOOC_2H_5$ (C), both with a concentration of 18 $mol\cdot m^{-3}$. NaOH (A) solution (900 $mol\cdot m^{-3}$) was slowly fed into the reactor through a feed tube with the inner diameter of 3 mm. The segregation index is used to quantify the product distribution and calculated from the following expression: $X_S = \overline{c_S}/\left(\overline{c_S}+\overline{c_R}\right)$, where $\overline{c_S}$ and $\overline{c_R}$ are the mean molar concentrations of species S and R after the reactions, respectively.

3.2 *Simulation strategy*

Numerical simulation was carried out by using the commercial CFD software Fluent. The whole tank was discretized as the computational domain. The virtual feeding pipe was created and the feed point coordinates are the same as experimental conditions. The Multiple Reference Frames model was applied to treat the rotational impeller. The flow field was obtained using the standard k-ε turbulence model in conjunction with the standard wall function. The SIMPLE algorithm was used to perform the pressure-velocity coupling. The second-order upwind scheme was adopted for the spatial discretization of all the transport equations.

In order to ensure the calculation accuracy and save the computational source, the preliminary work is the grid independence test. The final number of grids is about 680,000. The detailed simulation process was divided into the following four steps: (1) The turbulent flow-field was firstly solved with steady-state numerical simulation. (2) The transport equations for the mean mixture fraction and mixture fraction variance were solved as user-defined scalars, i.e., $X_1 = \langle \xi \rangle$,

$X_2 = \langle \xi'^2 \rangle$. The initial boundary conditions at the velocity inlet where reactant A was fed are $X_1 = 1.0$ and $X_2 = 0.0$, and for the computational domain inside the tank are $X_1 = X_2 = 0.0$. (3) The average values of turbulence kinetic energy ($\bar{k}$) and dissipation ($\bar{\varepsilon}$) in the reaction zone can be calculated according to Eq.(2-6). (4) The E-model (taking into the self-engulfment[14]) was introduced to calculate the segregation index, for the detailed steps please refer to the literature[15]. In this work, the continuous addition of reagent A was divided into $\sigma = 1000$ independent feed element to eliminate its effect on the prediction of X_S.

For a typical simulation work carried out in a single-phase stirred tank, it took about 24 h to calculate the flow field, and 1-1.5 h to solve the distributions of mixture fraction and mixture fraction variance. An additional 3-5 min was spent to compute the E-model.

4. Results and Discussion

4.1 *Depletion of NaOH solution in different conditions*

Figs.2 and 3 show the effect of agitation speed and feeding position on the depletion of NaOH solution (A) in the reaction zone after addition of the first feed element. When the impeller speed was increased and the feeding position was closer to the impeller, the consumption rate of the limiting reactant A increased due to the enhancement of mixing.

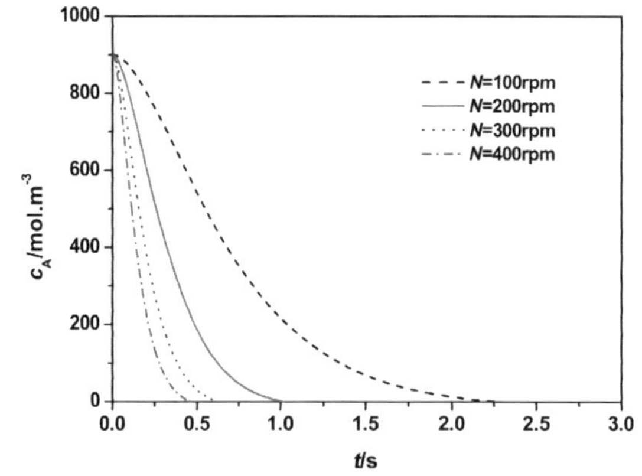

Fig.2 Depletion of the first portion of NaOH solution for different agitation speeds (Feeding position: F_{s2}, σ=1000)

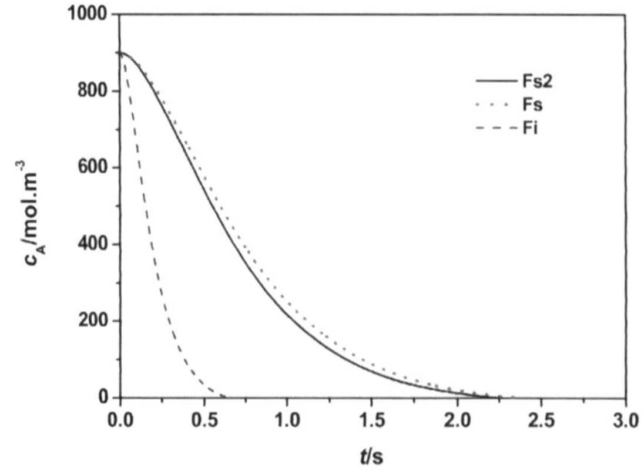

Fig.3 Depletion of the first portion of NaOH solution for different feeding positions (N=100 rpm, σ=1000)

4.2 *Effect of agitation speed on the segregation index*

Fig.4 shows the effect of agitation speed on the segregation index at different feeding positions. The simulation results are compared with the experimental data and the model-predicted values reported in literature[5]. As seen from Fig.4, no matter the feeding position near the liquid surface or the impeller, the segregation index decreases with the increase of agitation speed. In general, the new method developed in this work can capture the trend of segregation index, especially for the case that the feeding position near the surface. The largest deviation compared with the experimental data occurs at the feeing position Fi which is just above the impeller. This underestimate might be caused by the turbulence model and the modeling method of micro-mixing adopted in this work.

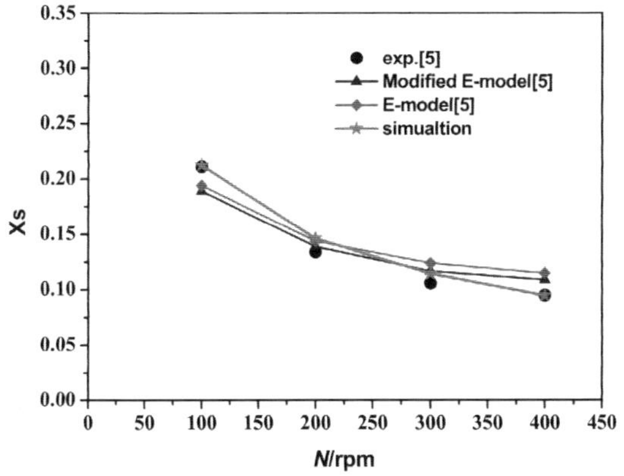

(a) Feed location: F_s

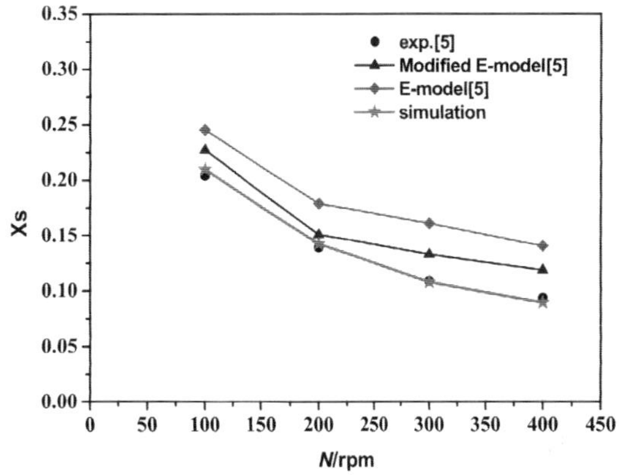

(b) Feed location: F_{s2}

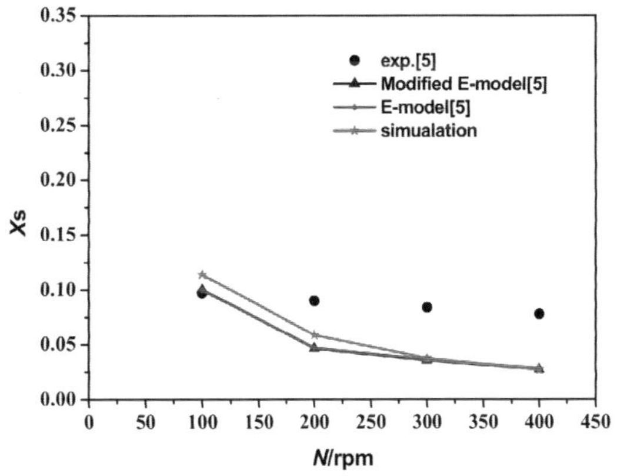

(c) Feed location: F_i

Fig.4 Effect of agitation speed on X_S for different feeding positions

4.3 *Distribution of mixture-fraction variance*

At the case of 100 rpm, Fig.5 shows the distribution of mixture-fraction variance in the y=0 plane for different feeding positions. The zones of $\langle \xi'^2 \rangle / \langle \xi'^2 \rangle_{max} \geq 0.001$ are marked in Fig.5. As it can be seen, the high values of mixture-fraction variance mainly concentrate on the zone close to the feed point. Due to the feed point locating at different positions in the stirred tank, the contour plots of the mixture-fraction variance can reflect the macroscopic flow characteristics and the difference of turbulent fields.

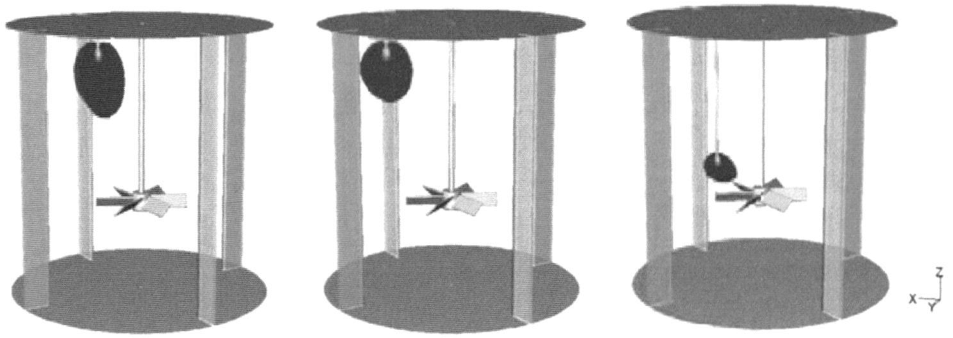

(a) Feed location: F_s (b) Feed location: F_{s2} (c) Feed location: F_i

Fig.5 $\left\langle \xi'^2 \right\rangle / \left\langle \xi'^2 \right\rangle_{max}$ distribution in the y=0 plane for different feeding positions

5. Conclusions

In this work, a CFD method combining with the micro-mixing E-model is proposed to predict the segregation index in a single-phase stirred tank. The key of this modeling approach is to get the turbulence kinetic energy and dissipation in the reaction zone through solving the transport equations of mixture fraction and its variance. The comparisons between the predicted segregation index achieved with our new numerical method and the experimental data are conducted. For the single-phase stirred tank, the model predictions of segregation index at different agitation speeds and feeding locations are in agreement with the experimental results, especially for the feeding point near the surface. In addition, all the simulation relatively cost less time. In the future work, it is worth to investigate micro-mixing with this method for industrial applications.

Acknowledgment

The financial supports from 973 project (2010CB630904), 863 project (2012AA061503), the National Natural Science Foundation of China (21306197) and the National Natural Science Fund for Distinguished Young Scholars (21025627) are gratefully acknowledged.

References

1. Cheng J C, Feng X, Cheng D, et al. Retrospect and Perspective of Micro-mixing Studies in Stirred Tanks [J]. Chin. J. Chem. Eng., 2012, 20(1): 178-190
2. Marchisio L, Barresi A A. CFD Simulation of mixing and reaction: the relevance of the micro-mixing model. Chem. Eng. Sci., 2003, 58(16): 3579-3587
3. Baldyga J, Henczka M, Makowski L. Effect of Mixing on Parallel Chemical Reactons in a continuous-flow stirred-tank Reactor. Chem. Eng. Res. Des., 2001, 79(A5): 895-900
4. Fox R O. Computational models for turbulent reacting flows [M]. Cambridge University Press, Cambridge, 2003
5. Akiti O, Armenante P M. Experimentally-validated micromixing-based CFD model for fed-batch stirred-tank reactors [J]. AIChE J., 2004, 50(3): 566-577
6. Han Y, Wang J J, Gu X P, et al. Numerical simulation on micromixing of viscous fluids in a stirred-tank reactor [J]. Chem. Eng. Sci., 2012, 74: 9-17
7. Baldyga J, Pohorecki R. Turbulent micromixing in chemical reactors--a review [J]. Chem. Eng. J., 1995, 58(2): 183-195
8. Fox R O. Computational models for turbulent reacting flows [M]. Cambridge University Press, Cambridge, 2003

9. Fox R O. CFD models for analysis and design of chemical reactors [J]. Adv. Chem. Eng., 2006, 31: 231
10. Vicum L, Ottiger S, Mazzotti M et al. Multi-scale modeling of a reactive mixing process in a semibatch stirred tank. Chem. Eng. Sci., 2004, 59(8-9): 1767-1781
11. Vicum L, Mazzotti M. Multi-scale modeling of a mixing-precipitation process in a semibatch stirred tank [J]. Chem. Eng. Sci., 2007, 62(13): 3513-3527
12. Gavi E, Marchisio D L, Baressi A A. CFD modelling and scale-up of confined impinging jet reactors [J]. Chem. Eng. Sci., 2007, 62(8): 2228–2241
13. Lui Y, Fox R O. CFD predictions for chemical processing in a confined impinging-jets reactor [J]. AIChE J., 2006, 52(2): 731–744
14. Bałdyga J, Bourne J R, Hearn S J. Interaction between chemical reactions and mixing on various scales [J]. Chem. Eng. Sci., 1997, 52(4): 457-466
15. Baldyga J, Bourne J R. The effect of micromixing on parallel reactions [J]. Chem. Eng. Sci., 1990, 45(4): 907-916

Abstract

"Effect of Bed-height and Number of Blade Passes on Powder Mixing Performance and Derivation of a Mathematical Relationship"

Siraj, M. S.[1,2*], Khinast, G. J.[1,3]

1. Graz University of Technology, Graz, Austria.
2. *Currently working at Pakistan Institute of Engineering & Applied Science (PIEAS), Nilore, Islamabad, Pakistan.
3. Research Center Pharmaceutical Engineering GmbH, Graz, Austria.

Solid particle mixing is an important industrial phenomenon and is crucial in chemical and pharmaceutical industries. In the latter case, it becomes critically important as it is directly linked to the drugs performance as the active pharmaceutical ingredient has to be mixed properly with the excipients in a drug in order to avoid blanks or super-potent pills. This leads to strict drug regulatory laws to be followed by the industry. These regulatory requirements and the Quality by Design (QbD) initiative in the process industry require a deeper understanding of the mixing phenomena on a basic level and the mixers being used for this purpose, which are in many cases convective mixers.

Some of the effects, e.g., blade angle and blade shape have already been studied [1-3]. In this work, however, a detailed investigation was done using a simple geometry [1,2] to study the effects of bed-height and number of blade passes on the powder mixing performance and more importantly with an objective to derive a mathematical relationship between these. Number of particles in this study were significantly higher as well as the length of the box than in our previous study [1]. Additionally, blade force and variation of inter-particle forces were also studied with the change in bed-height and number of blade passes. A mathematical relationship was also derived successfully and will be presented.

References

[1] M.S. Siraj, S. Radl, B.J. Glasser, J.G. Khinast, Effect of Blade Angle and Particle Size on Powder Mixing Performance in a Rectangular Box. Powder Technology 211 (2011) 100-113.

[2] M.S. Siraj, J.G. Khinast, Effects of Blade Shape and Angle on Solid Particle Mixing Using 2D and 3D Discrete Element Simulations, 544b (presented on Nov 10, 2010), 2010 AIChE Annual Meeting, Nov 7-12, 2010, Salt Lake City (UT) USA.

[3] G.R. Chandratilleke, A.B. Yu, R.L. Stewart, J. Bridgwater, Effects of blade-rake angle and gap on particle mixing in a cylindrical mixer, Powder Technology 193 (2009) 303–311.

Earlier – at the European Conferences on Mixing in London (2010) and Warsaw (2012), at the 2012 NAMF meeting in Mayan Riviera (Mexico), and in AIChE J 58 (2012) 3878-3890 – the results of 2 Direct Numerical Simulations (DNSs) were reported for a stirred vessel provided with baffles and a Rushton turbine, and operated under marginally turbulent conditions (Re=7,300). These simulations were of the Lattice Boltzmann type and were run on 128 IBM Power6, 4.7 GHz CPUs of the Dutch supercomputer SARA, the number of grid points used amounting to some $2{\cdot}10^9$. This spatial resolution implied that just in a very small domain in the near vicinity of the impeller blades grid cell size was not small enough to resolve the flow at the Kolmogorov scale. Some discrepancies were found for these DNS results when compared to experimental k and ε data available in the literature.

To further investigate these discrepancies , these Lattice Boltzmann DNSs were repeated on an INTEL platform of Xeon Phi processors for some finer grids. In this way, the effect of both grid resolution and Reynolds number can be investigated. During these runs, spectra are constructed on some strategically selected positions in the vessel. In this presentation, initial results will be presented.

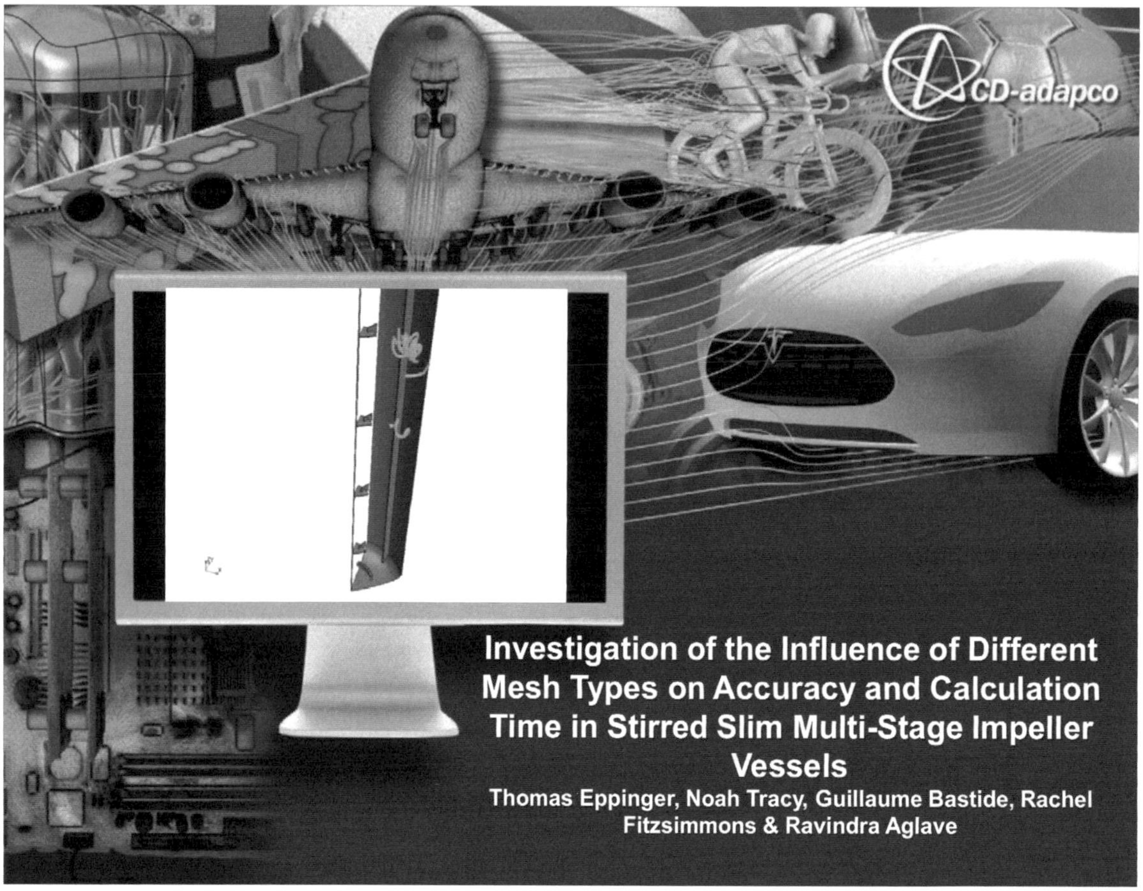

Agenda

CD-adapco

- **General Introduction & Motivation**
- **Setup**
  - Exp. Setup
  - Num. Setup
- **Results**
  - Mesh Study based on polyhedrons
  - Flow Pattern
  - Gas Holdup
  - Influence of different interaction forces
  - Simulation time
- **Summary**

18th Nov. 2014 AIChE Annual Meeting 2014, Atlanta, GA Thomas Eppinger

Introduction

CD-adapco

- **Larger Vessels needed for higher production rates**
- **Vessel diameter is limited**
  - transportation (place of manufacture is not place of operation)
  - Available space (when empowering existing plant)

→ Slim Vessels with H/T > 1

- **Design & scale up is difficult**
  - Intensive mixing requires multiple impellers
  - Various combinations possible (H/T, impeller types, rotational speed, baffles…)
  - Only few correlations available (compared to H/T = 1)
- **Design needs a lot of experiments → expensive**

→ CFD (constrain: Reasonably accurate results in a reasonable time)

Geometric Setup

CD-adapco

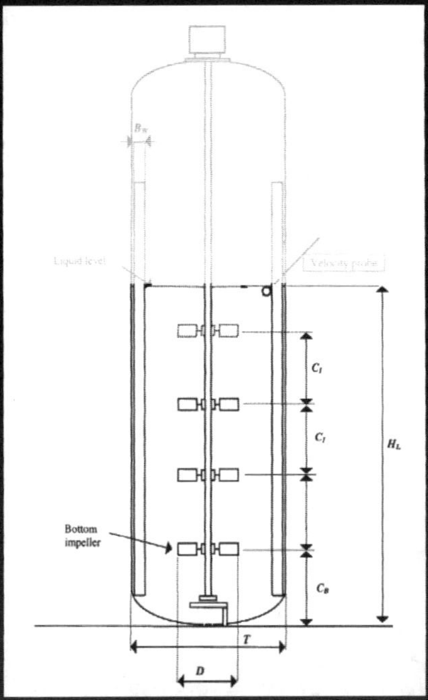

Property	Value
Rushton impellers	4
Blades per impeller	6
Blade height	0.14m
Blade length	0.17m
Bottom clearance C_b	1.12m
Impeller distance C_i	1.45m
Impeller diameter	0.7m
Liquid level H	6.55m
Liquid volume	22m^3
Tank diameter T	2.09m
Baffles	4

Vrabel, P. et al. (2000), Chem. Eng. Sci. 55

Setup in STAR-CCM+

CD-adapco

- CAD-Geometry
  - Parametric setup in built-in CAD-modeler
  - 60° segment & full model
    - Periodic BC
    - Add. baffles (total 6)
    - Negligible effect (Lane 2005)
- Steady Eulerian multiphase
- MRF & RBM
- Water / Air
- Kε-Turbulence model
- No-slip Wall BC
- Free surface as degassing boundary
  - Slip-wall for water
  - Permeable for air

Property	Value
Density Water	997 kg/m^3
Dyn. Viscosity Water	8.8871E-4 Pas
Density Air	1.18 kg/m^3
Dyn. Viscosity Air	1.85508E-5 Pas
Impeller Frequency	115 rpm
Vol. Flow Air	0.0263 m^3/s
Superficial Air Velocity	0.77 cm/s

Mesh Resolution

CD-adapco

- Mesh setup relative to one parameter: Base Size (40mm)
- Only exception: Prism layer thickness (constant = 4mm)
- Rotating region with higher resolution
- Sparger with higher resolution
- Conformal interface

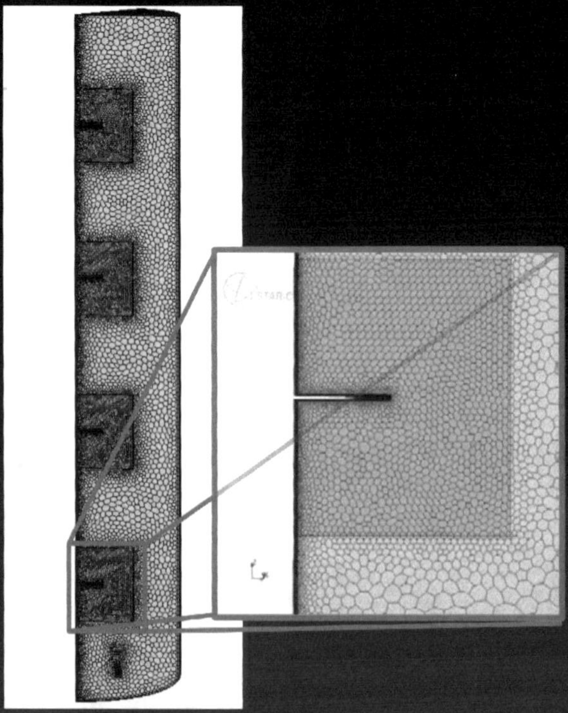

18th Nov. 2014 AIChE Annual Meeting 2014, Atlanta, GA Thomas Eppinger

Flow Pattern – Water & Gas Holdup

CD-adapco

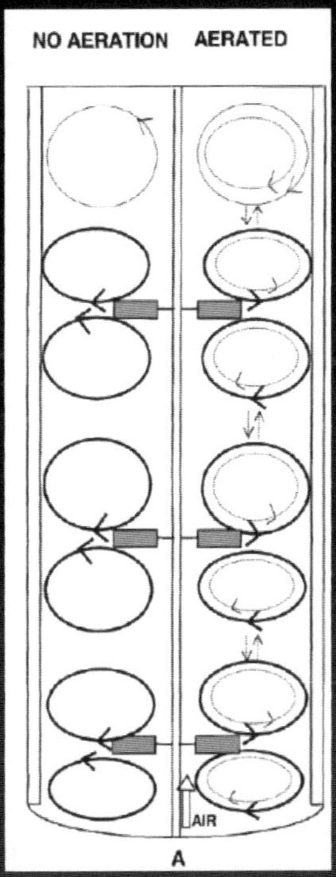

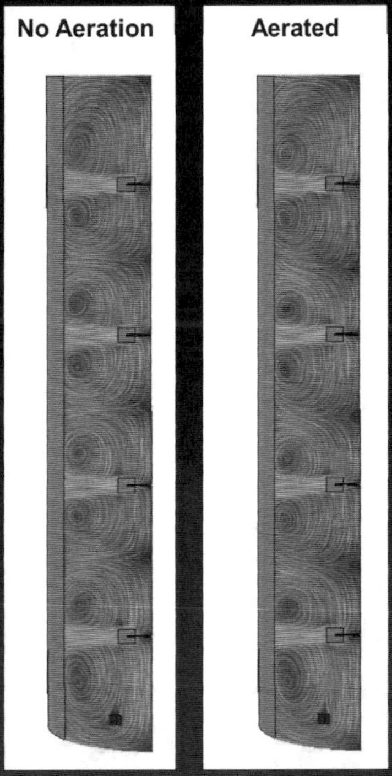

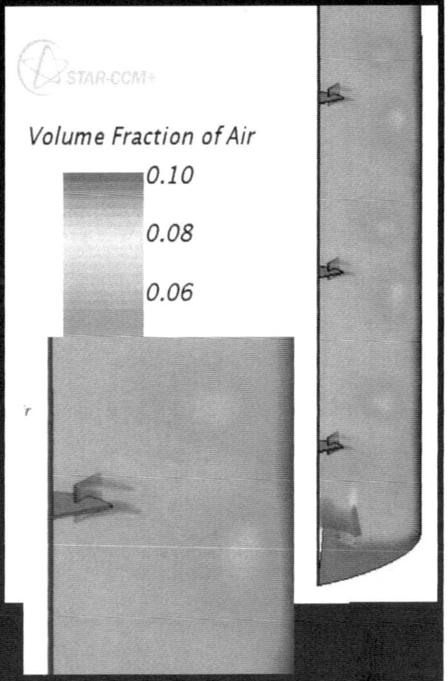

18th Nov. 2014 AIChE Annual Meeting 2014, Atlanta, GA Thomas Eppinger

Mesh Independency

CD-adapco

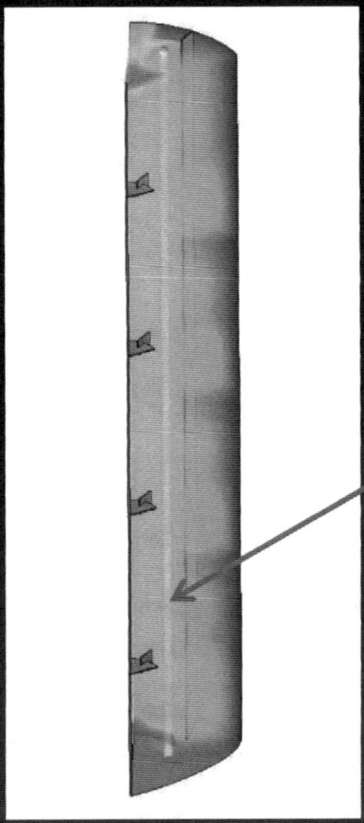

- Cell count approx.: 250k, 450k, 900k, 1.4M
- Polyhedrons
- Cell size relative base size, mesh is scaled except of the prism layer (constant thickness)
- Comparison based on line-probe with 200 data points
- Located in the center between impeller tip and baffle
- Monodisperse bubbles (3mm)

18th Nov. 2014 AIChE Annual Meeting 2014, Atlanta, GA Thomas Eppinger

Mesh Independency (Polyhedral Mesh)

CD-adapco

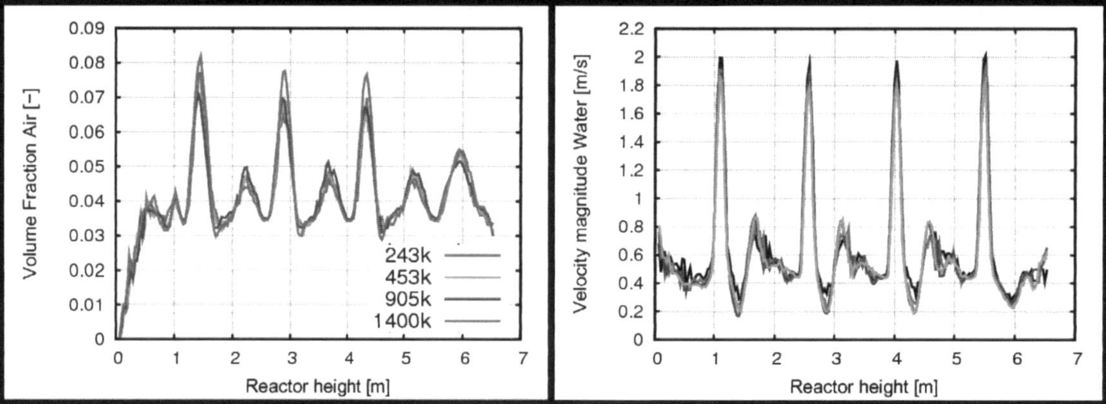

- Results are almost mesh independent even with coarsest mesh (243k cells)

18th Nov. 2014 AIChE Annual Meeting 2014, Atlanta, GA Thomas Eppinger

Influence of Bubble Size

CD-adapco

- Monodisperse bubble size (1, 2 and 3mm)
- 450k polyhedral cells
- S-gamma model incl. coalescence &breakup (log.-normal distribution: 1e-4mm < BS < 10mm)

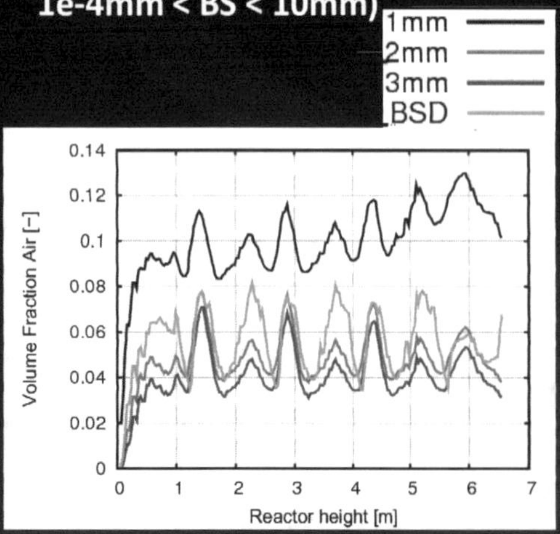

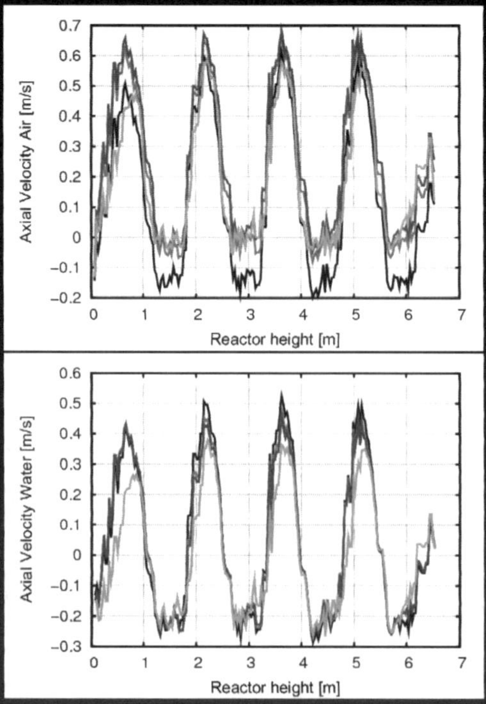

18th Nov. 2014 AIChE Annual Meeting 2014, Atlanta, GA Thomas Eppinger

Gas Holdup

- **Gas Holdup**
  - Exp. 4.7%
  - Size distribution 5.1%
  - Monodisperse 1mm ~9.5%
  - Monodisperse 2mm 3.7%
  - Monodisperse 3mm 3.5%

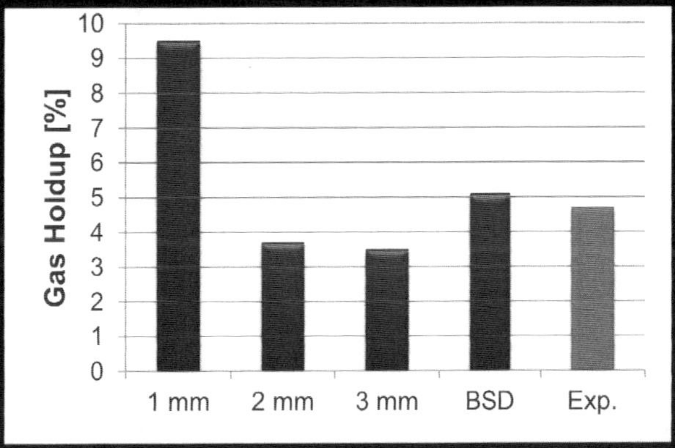

18th Nov. 2014 AIChE Annual Meeting 2014, Atlanta, GA Thomas Eppinger

Influence of Phase Interaction

- **Drag! (D)**
- **Buoyancy! (B)**
- **Turbulent Dispersion!**
- **Lift (LF)?**
- **Wall Lubrication (WLF)?**
- **Virtual Mass (VM)?**

- **Force influences always both phases : Fij = Fji**

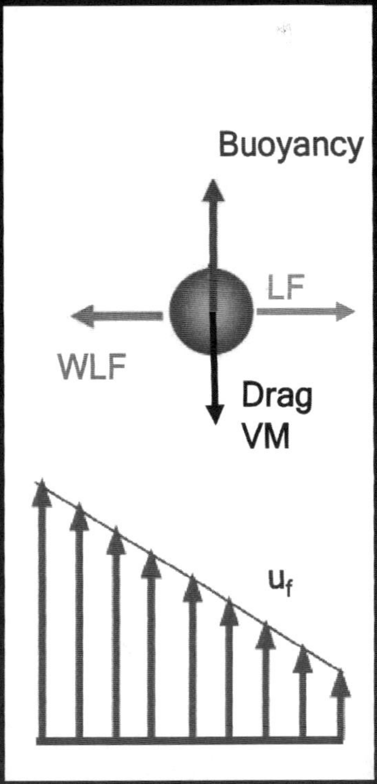

18th Nov. 2014 AIChE Annual Meeting 2014, Atlanta, GA Thomas Eppinger

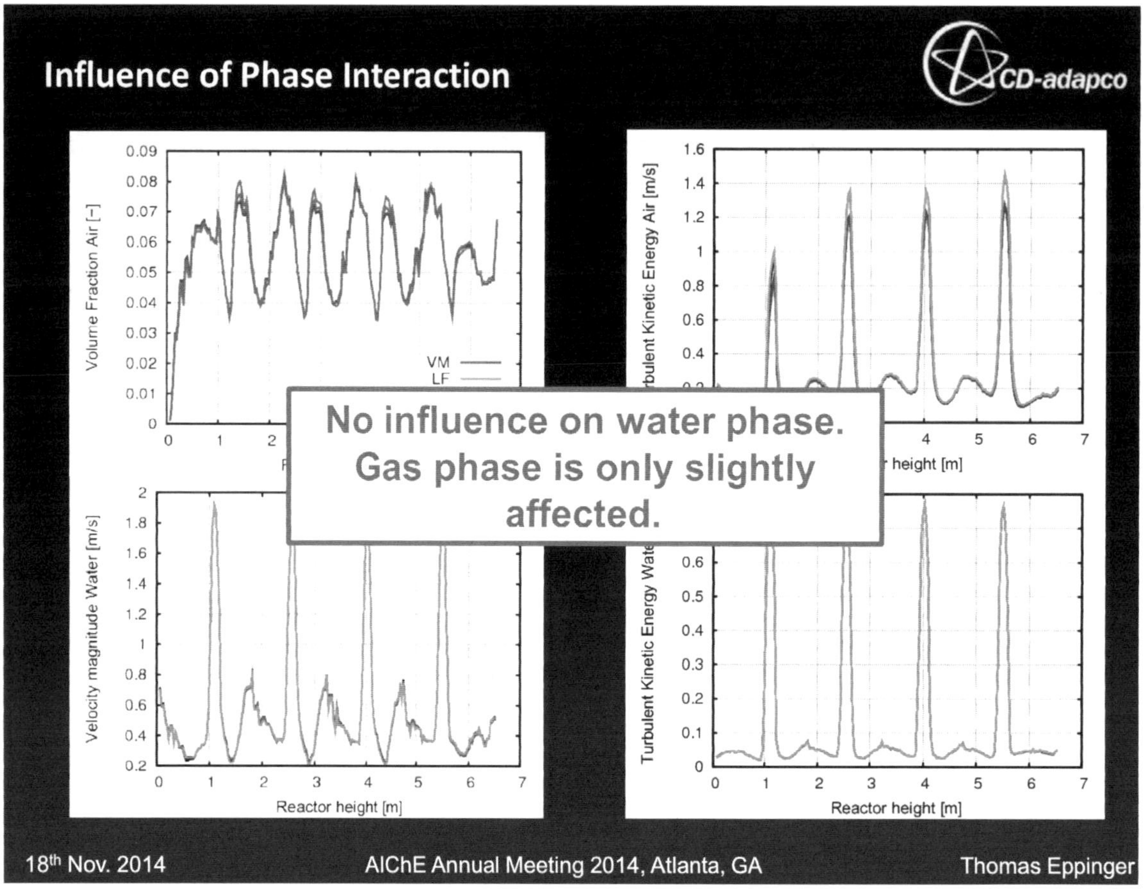
Influence of Phase Interaction
CD-adapco
Volume Fraction Air [-]
VM
LF
Turbulent Kinetic Energy Air [m/s]
No influence on water phase.
Gas phase is only slightly affected.
Velocity magnitude Water [m/s]
Turbulent Kinetic Energy Water [m/s]
Reactor height [m]
Reactor height [m]
18th Nov. 2014
AIChE Annual Meeting 2014, Atlanta, GA
Thomas Eppinger

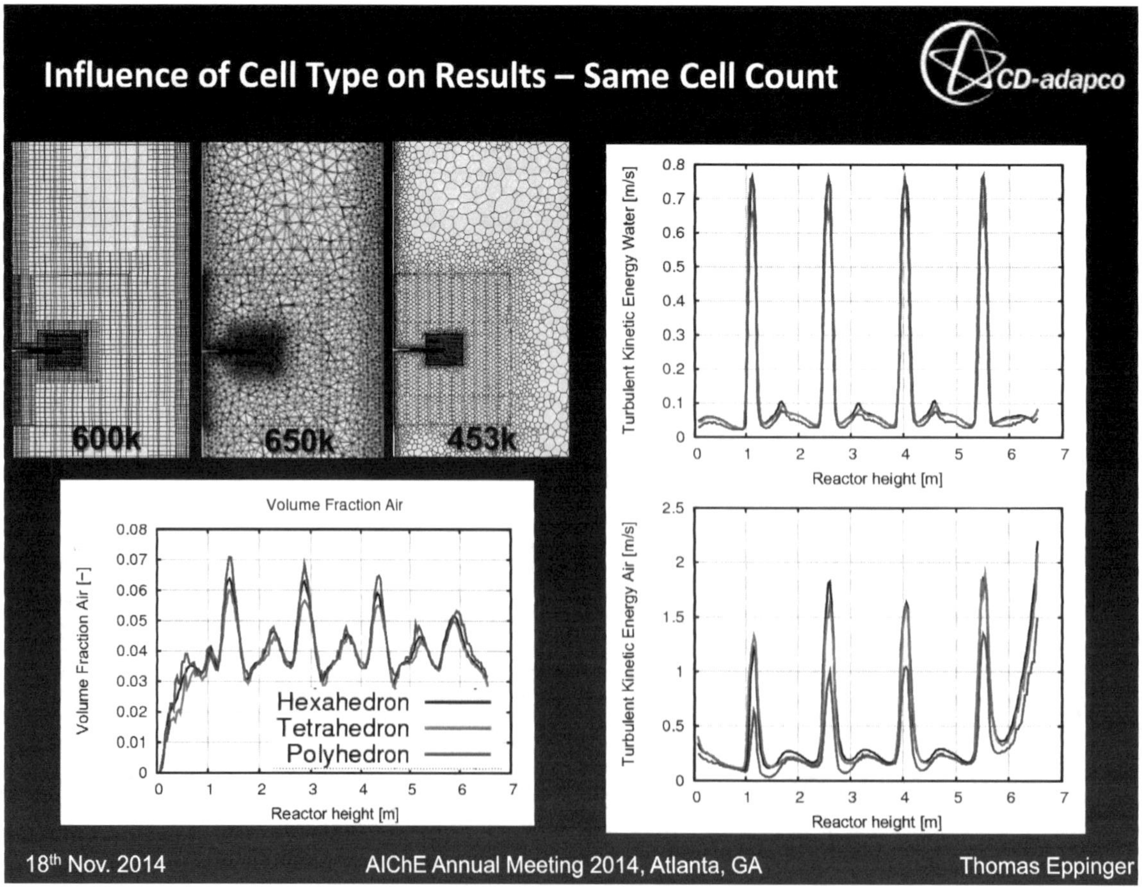
Influence of Cell Type on Results – Same Cell Count
CD-adapco
600k
650k
453k
Turbulent Kinetic Energy Water [m/s]
Reactor height [m]
Volume Fraction Air
Volume Fraction Air [-]
Hexahedron
Tetrahedron
Polyhedron
Reactor height [m]
Turbulent Kinetic Energy Air [m/s]
Reactor height [m]
18th Nov. 2014
AIChE Annual Meeting 2014, Atlanta, GA
Thomas Eppinger

Influence of Cell Type on Results – Refinement

CD-adapco

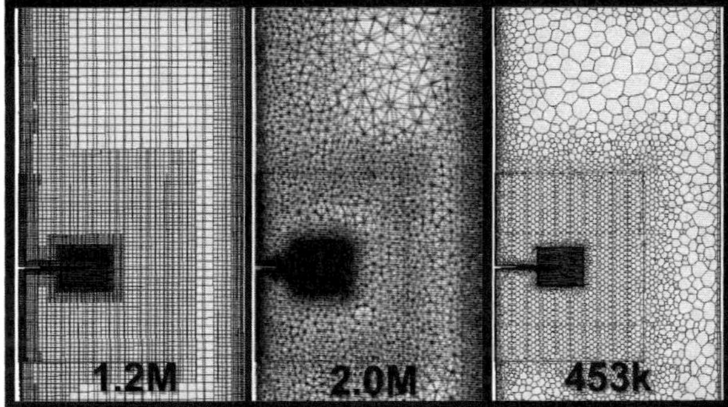

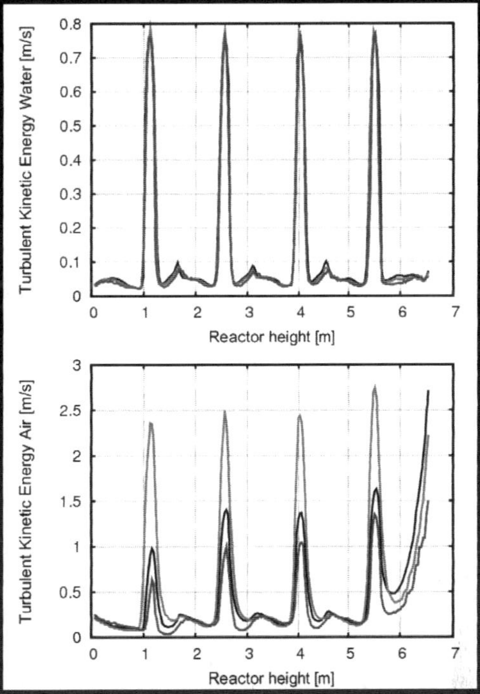

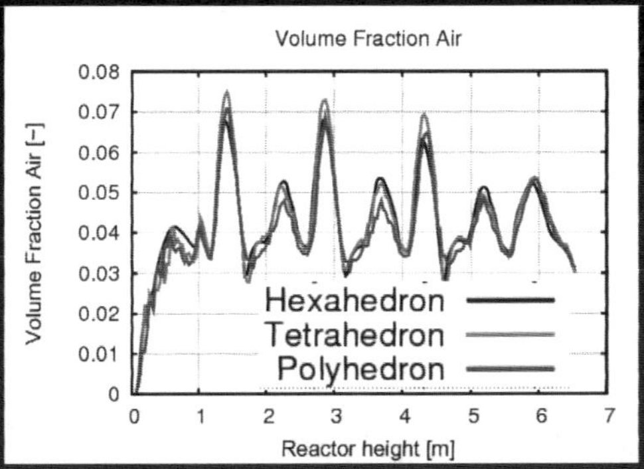

Influence of Cell Type on Simulation Time

CD-adapco

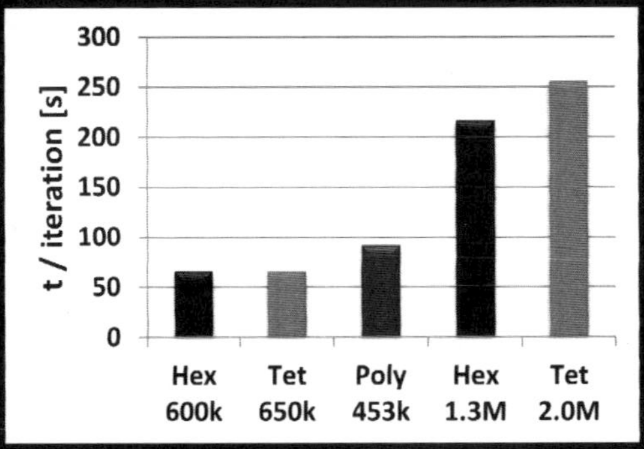

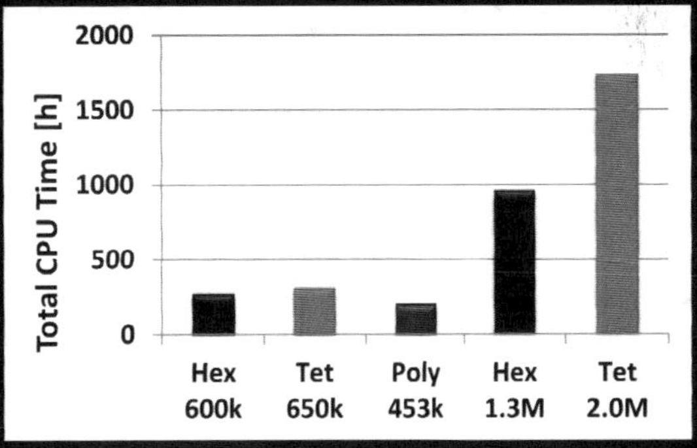

BUT

- Polyhedral cells need more time per iteration
- Convergence is much faster

Summary & Outlook

CD-adapco

- Virtual mass, lift force and wall lubrication force of negligible importance.
- Accurate results can be obtained with all cell types.
- Convergence is reached faster with a polyhedral mesh.

Thank you !

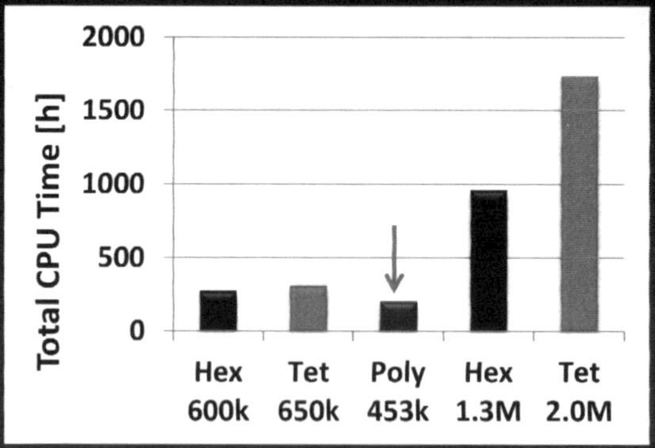

18th Nov. 2014 AIChE Annual Meeting 2014, Atlanta, GA Thomas Eppinger

393336 Jets Oscillation in a Long Horizontal Tank

Abstract:

In contrast to mixing in vertical tanks, jet mixing in long horizontal tanks is scarcely investigated in the literature. It is known that jet mixing in long horizontal tanks is more difficult when compared to short tanks, as more liquid volume must be recirculated through the jets.

In this study, computational fluid dynamics (CFD) simulations are conducted for the flow in a horizontal cylindrical tank with a length-to-diameter ratio of 5:1 and a liquid volume of 20,000 gallons. A pair of back-to-back Coldrey nozzles is placed near the center of the tank bottom, and each nozzle directs its jet towards the corresponding vessel end. An intriguing phenomenon is observed in the transient simulations, where the turbulent jets oscillate in both horizontal and vertical directions with a low frequency. In order to determine the source of such oscillation, a number of simulations are conducted to explore the effect of mesh type and size, boundary condition on the free surface, turbulence model, and time step. Oscillation persists in all cases, indicating that it is unlikely the result of some numerical instability. The oscillation also appears to be insensitive to the Reynolds number or symmetry in the nozzle or tank geometry. Another simulation with a single jet also shows the oscillatory flow behavior. Thus the oscillation is more likely to be caused by interaction between the jet and the recirculating flow in the tank, rather than interaction between the two jets. Further analysis of the jet velocity profile suggests that the secondary flow on the cross section of the jets might also contribute to the oscillation.

While similar confined jet oscillations have been previously observed in small cavities by both 2D numerical simulations and laboratory scale experiments, this study shows that such oscillation also exists in large industrial-scale horizontal tanks. The oscillatory motion of the liquid may lead to improved mixing in the tank.

354913 Modeling Mixing Processes Using Computational Fluid Dynamics and z-Transform

Abstract:

Computer simulations of mixing processes are typically performed at two levels: Computational fluid dynamics (CFD) can be employed to analyze the internal details of a mixing unit operation with high spatial and temporal fidelity, and flowsheet simulations can be used to study the overall performance of a mixing process as it interacts with other unit operations and responds to disturbances and control strategies.

Each of these simulation approaches has its own utilities as described above, but also its own limitations. The boundaries of a CFD model do not usually extended beyond the walls of the mixing process vessel, thus to a lesser or greater extent decoupling the process from the universe. In a flowsheet model the available transfer function description of a mixing process is often limited to prescribed black-boxes that are derived based on idealized assumptions, which may not be applicable to modeling complex processes or troubleshooting defective operations.

That said, CFD and flowsheet analysis are not mutually exclusive tools, and in fact they should complement each other quite well. In this presentation we will illustrate the fundamental aspects of using these two methods together by examining a linear system and its open-loop response, such as the residence time distribution (RTD) function of a continuous-flow stirred-tank reactor that might not necessarily exhibit classical first-order characteristics. We will show how to obtain a *z*-transform transfer function—the discrete equivalent of the Laplace transform transfer function—for the process from CFD simulation results. Then we will use the *z*-transform transfer function to generate additional RTDs corresponding to various conditions at the inlet to the mixing process, and compare these results with those obtained from the corresponding CFD simulations. To conclude our presentation, we will discuss issues related to the truncation error that is introduced when obtaining the *z*-transform transfer function from CFD simulation results.

373358 Simulating Fast Chemical Reactions and Mixing with Random Coalescence-Dispersion Modeling

Abstract:

A general code for simulation of turbulent flow and mixing with fast multiple chemical reactions using a combination of k-ε CFD and random coalescence-dispersion (C-D) is presented. Previously a code was developed to do approximate simulations of mixing and chemical reaction by utilizing C-D and a simple MATLAB simulation of the flows through volume segments of a stirred reactor vessel. Flows and rates of turbulence energy dissipation rates, hence mixing rates, were inferred from experimental data. The new code makes use of a k-ε CFD simulation of the flows and rates of turbulence energy dissipation and a previously published relationship between turbulence energy dissipation rate and random C-D rate. Since random C-D depends on representing the fluid by discrete fluid elements that flow and mix, mass rates of their movements from segment to segment must be equal to the simulated flow rates in order to maintain material balance.

The C-D method for simulating effects of mixing on multiple chemical reactions has the advantage that no closure hypothesis need be used. If the number of fluid elements per segment is high enough to be a statistically significant sample, then each element may be treated as a small batch reactor between each mixing event. C-D mixing events must be frequent enough to adequately simulate the progress of the mixing and chemical reactions. The method of simulation is tested by comparison of its results with experimental data.

Advantages and disadvantages of the C-D method as compared to other methods, such as complex reaction closures, large-eddy modeling, direct numerical simulation and lattice-Boltzman simulation, will be discussed. Example simulations using C-D for tubular and stirred-vessel reactions will be used to show how the method works.

377215 Supercritical and Subcritical Modes in Mixing in Rotating Microchannel

Abstract:

The flow and mixing in a rotating radial microchannel is investigated by both numerical simulations and experiments. Mixing is by crossflow due to Coriolis acting on the throughflow in the channel. For a given mixing quality from the two liquids in the channel, from the numerical simulation we found that there are two operating modes depending on the channel width-to-height (W/H) ratio. A smaller W/H corresponds to a subcritical mode for which the flow is dominated by viscous friction and mixing can be inhibited. On the other hand, a larger W/H ratio corresponds to operating under supercritical mode for which the time for liquid to flow from one side to the opposite side of the channel may become too long limiting the mixing. These two modes have been confirmed by experiments. We found that the supercritical has the advantage that a higher throughflow can be accommodated.

377321 Multiscale CFD-DEM Modeling of the Mixing of Viscous Liquids and Large Concentrations of Spherical Particles in Stirred Tanks

Abstract:

Despite the importance of laminar solid-liquid mixing operations in industry for the production of various goods such as pastes, greases, paints, pharmaceuticals and food products, a large proportion of the reported results in this field has been obtained in the turbulent regime. The intrinsic multiscale nature of solid-liquid flows in stirred tanks and the complexity arising from both the geometry of the impeller and the rheology of such suspensions, particularly in the case of high solids contents, create complex flow patterns within the tank, which may drastically affect both the yield and the quality of the desired product. Furthermore, the estimation of key mixing parameters such as the impeller torque, the just-suspension speed (N_{js}) and the cloud height remains unclear in all flow regimes. In particular, the Zwietering correlation has several limitations that may lead to poor predictions of N_{js}. Therefore, more modeling and experimental work is needed in order to shed light on these issues.

A variety of models have been developed to simulate solid-liquid flows. These include the classical Eulerian-Eulerian (or two-fluid) model and the combination of the Discrete Element Method (DEM) for the particles and CFD methods for the liquid (CFD-DEM). Despite their relative advantages and drawbacks, which will be briefly discussed during this talk, we believe that the CFD-DEM approach can be both efficient and accurate for the simulation of such flows. The objective of this work is then to introduce a multiscale CFD-DEM model, applicable to concentrated suspensions, wherein the particles are not resolved explicitly on the CFD grid but taken into account through the so-called volume-averaged Navier-Stokes equations. This entails projecting the particles onto the grid so that the solid-liquid momentum transfer term appearing in these equations, such as the drag force exerted by the particles on the liquid, can be locally evaluated. After describing the proposed model in detail, simulation results obtained with it for the laminar mixing of concentrated solid-liquid suspensions in stirred tanks will be presented.

383196 Applications of Mean Age Theory to Mixing in Wastewater Treatment Facilities

Abstract:

Mixing operation is essential to wastewater treatment facilities. In most such facilities efficient mixing is required to homogenize mixtures, maximize contacts between constituents, prevent solid settling, etc. Different types of mixing devices are used to achieve these objectives, such as agitated impellers, jet mixers, floating mixers, static mixers, and other custom designed mixing systems. The design and selection of a mixing system has been mainly based on past experience and a quantitative method has been lacking. More recently, CFD modeling techniques have been used to facilitate the design process but most of such applications have been limited to fluid flow patterns.

In this talk, a quantitative method using CFD tools to design a mixing system will be discussed. The method is based on the 3D solution of mean age distribution. This method can be used to quantitatively measure either a batch system or a continuous system. The spatial distribution of mean age reveals the non-uniform distribution of local blend time thus any bypassing or dead zones can be easily located and quantified. The statistical analysis of mean age distribution provides quantitative comparisons of mixing efficiency of different mixing designs and guides the selection of an optimal design. Several examples of applying mean age method to mixing problems in wastewater treatment will be discussed in the talk.

CFD Simulation of Mixing in Wastewater Treatment

Binxin Wu

Philadelphia Mixing Solutions Ltd., Palmyra, PA 17078

Abstract

Aerated lagoons and anaerobic digesters have been widely used to treat wastewater. In this study, a computational fluid dynamics technique is used to simulate mixing in aerated lagoons and anaerobic digesters, in which the rotation of a mechanical impeller is modeled using multiple reference frame method and wastewater is assumed to be a Newtonian or non-Newtonian fluid depending on total solids (TS) concentration. Numerical simulations of mixing in a small lagoon (173.74 m × 129.54 m × 3.66 m) with 4 aerators and in a large lagoon (471.77 m × 129.54 m × 3.66 m) with 12 aerators are conducted, and quantitative comparisons of the average velocity and biological oxygen demand versus placement of aerators in each lagoon are made, assuming wastewater to be water. In a digestion tank with a diameter of 20 m and a liquid height of 10 m, mixing by two submersible mixers is simulated, in which the substrate is assumed to be a pseudoplastic or Herschel-Bulkley non-Newtonian fluid with TS ranging from 2.5% to 12.1%. The results show that the mixing energy level increases nonlinearly with the increase of TS to achieve the same average velocity of the flow field.

Keywords: Bioreactors, Mixing, Computational fluid dynamics, non-Newtonian fluid, biological oxygen demand.

382148 CFD for the Study of the Multiphase Flow and Mass Transfer in a Stirred Tank Reactor for the Screening of Shaped Catalysts

Abstract:

Many reactions in the petrochemical, refining and biofuel industries occur in the presence of shaped solid catalysts, in the form of beads or extrudates. These particles have typical dimensions of the order of 1 to 3 mm, catalysing reactions occurring in liquids or gas-liquid mixtures. Before the selection of a catalyst for an industrial process, primary screening tests need to be performed previously in laboratory-scale reactors to assess the catalyst efficiency in its commercial shape, preserving the spatial distribution of active sites. These catalyst testing reactors need to fulfil some essential requirements to allow for a correct assessment of the kinetics of the reactions [1]: isothermal operation; appropriate flow distribution and mixing rates; and the absence of interphase mass transfer limitations. If these requirements are not met, the measured rate will not be equal to the reaction rate but instead to the rate of the limiting transport phenomenon, i.e., mixing or mass transfer. With the continuous development of new and more efficient catalysts, and the growing interest in the use of complex fluids like heavy crude feedstock or biomass-derived, catalysts testing and selection in lab-scale reactors become often hindered by interphase mass transport limitations or by hydrodynamics.

One of the existing types of reactors used for catalyst testing are stirred tank reactors with annular stationary catalytic baskets, also known as Robinson-Mahoney (RM) catalyst testing reactors [2]. In RM reactors, the shaped catalyst is introduced in an annular basket made from a metallic net or a perforated metal sheet, and immersed in the liquid. An axial flow impeller is then placed in the center of the annular basket to promote simultaneously the flow through the catalyst walls and particles, and the mixing of the bulk. Although RM reactors have been reported to offer less mass transfer limitations comparatively to laboratory fixed bed reactors [3], its way of operation is not yet completely understood and mass transfer limitations can still occur. Small changes in catalytic basket reactor operations (rotation speed, liquid viscosity, foam formation,...) may greatly reduce the reactor performance, and recommend measuring mass transfer coefficients for the reaction conditions before performing the kinetic measurements [3]. In addition, geometrical changes to the basket or impeller designs, or the selection of a different catalyst filling, of different shape or size, can also lead to different reactor performances, even when similar operating conditions are chosen. Measurements of mass transfer rates on RM reactors are not abundant in the literature, and are always specific for reactor geometry, working fluid and catalyst shape and size [4-6]. This happens because the extensive experimental study of RM reactors at the laboratory becomes neither practical nor economical when a large number of geometrical or operational parameters needs to be studied.

In this work, Computational Fluid Dynamics (CFD) is used to simulate the gas-liquid flow in a commercial lab-scale RM reactor. A multiphase 3D CFD model of gas-liquid flow in the reactor with the stationary catalytic basket has been developed with ANSYS Fluent. The studied reactor is a cylindrical, flat-bottomed tank with a diameter of 58 mm and a total volume of 300 cm^3, from *Top Industrie*, designed for the kinetic testing of shaped catalysts. Mixing is promoted by a stainless steel radial flow self-inducing impeller, which simultaneously promotes de induction and dispersion of the gas phase in the liquid. A static annular basket is placed inside the tank around the impeller to support the shaped catalyst particles. The basket walls are made from a perforated steel sheet with a thickness of 0.5 mm and perforations with a diameter of 0.8 mm, corresponding to an open area of 45%. The radial flow turbine promotes the flow across the basket annular section, around the catalyst particles, from the impeller site until the tank wall. Four baffles with a width of 10 mm and a thickness of 1 mm are placed above the basket. The Eulerian-Eulerian multiphase flow model was chosen for the simulation of the gas-liquid flow. Turbulence in the flow, exteriorly to the basket, was simulated with the Standard Multiphase k-εModel, and the Multiple Reference Frame (MRF) approach was chosen to deal the impeller motion. The flow in the basket filling was simulated as a homogeneous and unconsolidated porous medium with the Brinkman-Forchheimer equations, and the perforated basket walls were simulated as semi-permeable zero-thickness surfaces with a pressure drop. The numerical results were compared with experimental data from Particle Image Velocimetry, a bubble tracking imaging technique and gas absorption measurements.

The CFD simulations allow knowing the velocity field and gas volume fraction in each point of the domain. This information is used to estimate the mass transfer coefficients between the solid-liquid and gas-liquid phases, k_S and $k_L a$, respectively, and compare it to the characteristic time of the catalysed reaction. Simulations for different operational conditions are used to identify the operational regimes and the range of conditions that allow a correct assessment of reaction rates with the existing setup. The developed CFD flow model will be used as well in the future as an optimisation tool for the design of enhanced reactor configurations.

References

[1] C. Perego, S. Peratello, Experimental methods in catalytic kinetics, Catalysis Today, 52 (1999) 133-145.

[2] J.A. Mahoney, K.K. Robinson, E.C. Myers, Catalyst evaluation with the gradientless reactor, Chemtech, 8 (N12) (1978) 758-763.

[3] I. Pitault, P. Fongarland, M. Mitrovic, D. Ronze, M. Forissier, Choice of laboratory scale reactors for HDT kinetic studies or catalyst tests, Catalysis Today, 98 (2004) 31-42.

[4] M. Mitrovic, I. Pitault, M. Forissier, S. Simoens, D. Ronze, Liquid-solid mass transfer in a three-phase stationary catalytic basket reactor, AIChE Journal, 51 (2005) 1747-1757.

[5] I. Pitault, P. Fongarland, D. Koepke, M. Mitrovic, D. Ronze, M. Forissier, Gas–liquid and liquid–solid mass transfers in two types of stationary catalytic basket laboratory reactor, Chemical Engineering Science, 60 (2005) 6240-6253.

[6] M. Braga, Étude des phénomènes de transfert et de l'hydrodynamique dans des réacteurs agités à panier catalytique, in: Ph.D. Dissertation, Claude Bernard University, Lyon, 2013.

386338 Steady State CFD for Determining Just Suspended Speed: Theory & Results

Abstract:

A new method for determining the just suspended speed (N_{jss}) in stirred tank reactors was previously developed using steady state CFD simulations. While preliminary results were somewhat empirical based, the method has been improved with more rigorous theoretical backing that is analogously aligned with mean age theory.

Scalar transport equations were solved to track the solid phase independently from the bulk fluid. In this way, N_{jss} was determined based on the mean age of particles in a settling zone along the bottom of the tank. Plotting average age for the solid phase in this region versus impeller speed reveals two distinct magnitudes, with the transition corresponding to N_{jss}. Above N_{jss}, the average age is low due to large convective velocities. Below N_{jss}, the convective scalar transport is negligible, so the average age is several orders of magnitude higher due to the very low rate of diffusive scalar transport.

Simulations have been performed for a range of solids concentrations and various tank geometries. Computational results compare very favorably with published experimental data. This method can be used for determining N_{jss} for systems in which it would be difficult with traditional experimental techniques. It also removes the subjectivity associated with traditional experimental visualization techniques. Most importantly, since the technique employs steady state simulations, it is advantageous over time-dependent simulations, which require much longer computational times due to the multiple phases and transient nature.

393903 Minimum Speed for Off-Bottom Solids Suspension: Zwietering or P/V

Abstract:

Experiments were made to study the effects of solids suspension parameters on correlations such as Zwietering and a constant P/mass approach. Tests were made in 3 geometrically similar tanks with tank diameters of 0.3m, 0.5m, and 1.0m with various axial flow impellers. CFDs of each condition were also made to determine what the conditions were near the locations of the last suspended particles.

There are many articles published about solids suspension in mixed tanks. Many are confusing, and many contradict each other. We decided to launch an experimental and computational study to better understand how disagreements between researched results can happen. The most commonly used correlation for scale-up is based on the work of Zwietering. But many people have criticized it. More recently, there have been claims that the scale-up should be based on constant P/V, and others that it should be a hybrid between the two models.

Our study looks at 12", 20" and 40" tanks and a variety of hydrofoil impeller designs and pitched bladed turbines of various impeller to tank diameter ratios and various off bottom clearances. We varied the solids concentration from 1 to 65%. We varied the particle sizes from 80 to 1000 microns. Every experiment was duplicated with CFD to better understand the hydro-dynamics along the tank base and the walls. Based on our results we plan on expanding our work to a 96" tank and investigate the effect of solids as small as 20 microns on solids suspension.

The impeller design does have an impact on solids suspension, but we also found that for geometrically similar setups, solid suspension was fundamentally different between the 12", 20" and 40" tanks. We also discovered a "sweet" spot for each impeller design, where the power of suspension was at a minimum for that tank geometry and slurry characteristics. An amazingly simple little "trick" will be demonstrated to improve solids suspension in many full-scale mining applications. Finally, we will also show how constant P/V and Zwietering can both be correct scale-up methods.

381029 Mixing Efficiency of Longitudinal Muscle and Segmentive Contractions in the Small Intestine

Abstract:

It is largely recognize in literature that muscular contractions of the small intestine improve nutrient absorption by facilitating the mixing of intestinal contents during digestion and the continuous renewal of nutrients at the intestinal wall.

Two specific motilities which have been observed in the rabbit during ex-vivo experiments are longitudinal motions and segmentive contractions. Longitudinal motions act to move the organ axially with no change in the diameter of the intestine. Segmentive contractions act locally to partially occlude the intestinal core.

The goal of this work was to numerically investigate how these two different types of contraction affect the dynamics and mixing of intestinal fluids of different viscosities. The rabbit intestine was modelled as an elastic solid (Young's modulus = 3000 Pa, Poisson's ratio = 0.35) and was simplified to be of cylindrical geometry. The motion of the intestinal wall was imposed based upon video data from ex-vivo observations of the rabbit ileum.

The governing system of mathematical equations were solved in the CFD software package Polyflow (ANSYS). The behaviour and stretching of two Newtonian fluids (10-3 Pa.s and 1 Pa.s) and one non-Newtonian fluid (of Bird-Carreau type with exponent 0.41, zero shear viscosity 0.65 Pa.s, and infinite shear viscosity 0.501 x10-8 Pa.s) were numerically analysed. The mixing efficiency was quantified using an in-built post-processor by measuring the stretching and folding of differential fluid elements.

It was found that the shear rates that develop as a result of these motions are too small to cause significant pseudoplastic behaviour of the fluid under study. Stretching rates were found to be dependent on both the properties of the fluid and the contraction type. High rates of stretching were observed along the entire wall during a longitudinal motion, but the dynamics suggest that flow is mostly axial with little mixing between the core and wall. The stretching of fluid elements was much more localised during a segmentive contraction, but more movement between the core and the wall was observed. Viscosity was shown to be an important factor with much higher stretching rates in the low viscous fluid. The non-Newtonian fluid behaved similarly to a viscous-Newtonian fluid, due to the low shear experienced.

Simulations also showed that blocking one end of the intestine increased the rate of element stretching during a segmentive contraction, with the rate of stretching increasing the closer to the blockage the contraction occurred.

Results from this study illustrated the effect of different intestinal motility patterns of the wall on the dynamics and mixing efficiency of intestinal contents during digestion

376470 The Use of CFD Modeling to Understand Variability in Particle Agglomeration in an API Crystallization

Abstract:

Powder property control for Active Pharmaceutical Ingredients (API) is important to ensure efficacy and to enable consistent manufacturing of the drug product. These properties can be controlled through crystallization design, targeting conditions that favor particle growth relative to nucleation and agglomeration. During API crystallization development, the process parameters critical to ensure particle growth are identified, addressing the challenges in scaling from the laboratory to the plant. Here, an API crystallization is discussed where no single mode – nucleation, growth, or agglomeration – is dominant. Initial process scale-up identified a mixing dependence resulting in variable agglomerate formation. Standard mixing correlations (tip speed, energy input) did not provide predictive agglomerate control.

These scale-up results prompted the development team to seek a more predictive mixing model to provide greater insight into agglomerate formation and control. MixIT, a software developed by Tridiagonal Solutions, was used to perform computational fluid dynamics (CFD) analysis of mixing conditions across the pilot scale reactors. For each reactor under consideration, a mesh was generated and the 3D, steady-state Navier-Stokes equations were iteratively solved. As an output of the analysis, important mixing parameters such as tip speed, average strain rate, and impeller power number were predicted. Further, distribution curves (e.g. volume percent vs. strain rate) provided a richer understanding of the strain rate experienced by particles in different pilot scale reactors. These curves provided insight into the different mixing conditions which may influence agglomerate formation. The mixing simulations were then applied to the proposed manufacturing scale crystallizer and process parameters were selected to match the desired strain rate distribution. The resulting batches demonstrated consistent powder property control in the larger scale API crystallizer.

381219 Large Bioreactor Simulation with Graphic Processors

Abstract:

So far, the engineering process of the reactor design was mostly driven by empirical knowledge. The effectiveness of the bioreactor and thus the biopharmaceutical manufacturing process could be enhanced by combining the state of the art empirical knowledge with a multiphase and multiscale simulation. Simulation results could then be used for the development and the design of new reactor generations. However, despite the recent improvement of computational capabilities, only a few seconds of real time operation of an industrial scale reactor takes months to simulate..

Thus, the goal of this study is to use graphic cards to speed up this simulation. The Compute Unified Device Architecture (CUDA) technology of nVidia has made the computational power of graphic processing units (GPUs) available for scientific calculations [1]. To take full advantage of the processor's parallelism, localized calculation algorithms are needed. Therefore the lattice Boltzmann method is used for modelling the fluid flow field, which is agitated by a Rushton turbine. The method is based on the lattice gas automata [2] and uses a regular grid with evenly distributed nodes. The static and moving boundaries are modelled with the modified bounce back algorithm.

The calculation of the bubble movement is done by solving the Newton's equation of motion. The sum of the forces acting on each bubble, i.e. the drag, the buoyancy, the lift force, the history force, the added mass effect and gravity, is used to determine the acceleration of the particle. The acceleration and the time step length give the velocity and the position change at the end of the time step. For coalescence and breakup stochastic models are used. The phases are coupled with a two way approach.

As the simulation results are produced on the graphic card, intermediate data can be visualized while the calculation is running. To simulate large reactors, the code has a multi GPU functionality, hence it can distribute the workload of the simulation on several graphic processors.

Validation is done with literature data as well as with results from holdup measurements of a 150l laboratory reactor.

References

[1] http://www.nvidia.com/cuda

[2] Sukop, M.: Lattice Boltzmann Modeling, Springer, Berlin, 2007.

383506 On the Use of Periodically Inverted Assay Tubes As a Screening Platform for Recombinant Escherichia coli Cultures

Abstract:

The screening of strains and culture conditions is a routine operation in biotechnology settings. Here we illustrate the use of commercial Vacutainer™ assay tubes agitated by successive periodic inversions as an alternative platform to conduct early development culture experiments of two recombinant strains of *Escherichia coli*. As a model system, we use two *E. coli* strains of biopharmaceutical relevance: (a) a C41 derived strain producer of a monoclonal antibody fragment (Fab) and a C41 strain producer of a dimer of the globular region of an influenza antigen. Tubes (effective volume of 3 mL) were gently agitated by periodic inversions in a commercial rotational mixing system (ATR Rotamix, USA). Experiments were conducted using three growth media formulations for *E. coli* with and without replacement of the air in the tube head space. We compared the culture performance of the periodically inverting assay tube mixing system (pITMS for short) with that of shake flasks (150 mL of effective volume) and fully instrumented bioreactors (1.3 L of effective volume) for the same strains and culture conditions. Results produced in the pITMS exhibit high reproducibility and low variance among different culture experiments. Even though the growth rate (μ in h^{-1}) was different in each type of reactor, the general growth behavior of the bacterial culture and the levels of protein expression in assay tubes was practically the same as in the shake flasks and in bioreactor, demonstrating the predictive potential of the pITMS.

Our results suggest that the pITMS is a practical alternative to conduct parallel screening experiments that can resemble the behavior of aerobic bacterial cultures performed in flasks and instrumented bioreactors.

387636 Characterization of Oxygen Availability Measured As Oxygen Transfer RATE and Dissolved Oxygen Tension in Shake Flasks for the Production of O-Mannosylated Proteins in Streptomyces Lividans

Abstract:

Introduction. Shake flask are the most frequently used reaction vessels in biotechnology (Buchs, 2001). In spite of their large practical importance, very little is known about the characteristic properties of shaken cultures from an engineering point of view. Oxygen transfer rate (OTR) is one of the most suitable measurable parameter to quantify the physiological state of an aerobic culture, since most metabolic activities depend on oxygen consumption (Buchs, 2001). The effect of three different shear and aeration conditions using conventional normal (NF), baffled (BF) and coiled (CF) shake flasks on *Streptomyces lividans* cultures on OTR was evaluated online. Also, the dissolved oxygen tension (DOT) was measured on the three flasks designs. Previous reports of our group indicates that different flask designs in *S. lividans* cultures leads to changes in *O*-mannosylation and production of a recombinant protein (Gamboa-Suasnavart et al., 2011), suggesting that oxygen transfer rates are also affected.

Methods. *S.lividans* cultures were performed at 30ºC, 150 rpm, 60 h in 34% sucrose Luria- Bertani medium (Gamboa-Suasnavart et al., 2011). *Corynobacterium glutamicum* cultures in same conditions were carried as a higher respiration activity microorganism for comparison. OTR was measured using the RAMOS (Anderlei and Buchs, 2004). This device analyzes the depletion of oxygen concentration in the gas headspace of the shake flask with an oxygen sensor. The slope of this depletion corresponds to the OTR (Anderlei and Buchs, 2004). Carbon transfer rate (CTR) was calculated as the carbon dioxide formation, measured with a sensor in the flask headspace. Three coiled, three baffled and two conventional flasks were used each time. RQ was calculated as the ratio between CTR/OTR. DOT was measured by a luminescent patch (Pst3) using the system Fibox3 by PreSens.

Results and Discussion. Cultures carried out in CF and BF reached a maximum OTR of 9.16 ± 0.15 and 9.36 ± 0.28 mmol/lh respectively after 40 hrs of culture. In contrast, a significantly lower OTR was reached (0.66 ± 0.48 mmol/lh) in those cultures in NF. *C. glutamicum* cultures presented the same trend, achieving in CF, BF, and NF, 10 ± 0.22, 9 ± 0.31 and 2 ± 0.17 mmol/lh, respectively. RQ values was closer to one in BF and CF, indicating an efficient aerobic metabolism. But in NF, RQ values fluctuates between one and two, suggesting an inefficient carbon oxidation and probably used in other metabolism pathways.

DOT in BF and CF decrease up to 50% at the end of the culture but in NF a total depletion of oxygen was observed. This might be due to the poor OTR observed in NF cultures compared to BF and CF. On the other hand, the pellet sizes of *S. lividans* in NF are seven times (1.57 ± 0.41) higher than those in BF and CF (0.23 ± 0.06 and 0.16 ± 0.05) (Gamboa-S et al. 2011). By using the equation reported by Cui et al. (1997), an oxygen transfer limitation can be assumed by the higher dissolved oxygen tension needed to diffuse to the center of the pellet (170% DOT). In the case of the pellets from BF and CF, just 1.76% of DOT is required, demonstrating that in those geometries no oxygen limitations can occur.

Conclusions. Flask design modifies the OTR in microorganism cultures with low and high respiration activity. Higher OTR values are reached when baffles or a spring were used, due to the increase in mass transfer. OTR changes lead to a differences in metabolism as is shown in RQ values. DOT in conventional flask is insufficient to achieve a properly diffusion into pellets in almost whole culture. These OTR differences could lead to changes in recombinant glycoprotein production in *S. lividans* cultures. The knowledge about the oxygen transfer aspects inside flasks can provide new tools for the development and scale-up of biotechnology processes, mainly in those producing bacterial recombinant glycoproteins.

Acknowledgements. CONACYT-INNOVAPYME 181895, CONACYT 178528, 104951-Z y PAPPIT-UNAM IN-209113, IN-210013

References.

1. Büchs J, **Introduction to advantages and problems of shaken cultures**, *Biochem Eng J*, 2001, **7**:91-98.

2. Anderlei T, Büchs J, **Device for sterile online measurement of the oxygen transfer rate in shaking flasks**, *Biochem Eng J*, 2001, **2**:157-162.

3. Gamboa-Suasnavart RA, Valdez-Cruz NA, Córdova-Dávalos LE, Martínez-Sotelo JA, Servín-González L, Espitia C, Trujillo-Roldán MA, **The *O*-mannosylation and production of recombinant APA (45/47 KDa) protein from *Mycobacterium tuberculosis* in *Streptomyces lividans* is affected by culture conditions in shake flask**, *Microbial Cell Factories*, 2011, 10:110

4. Cui YQ, van der Lans RG, Luyben KC, **Effects of dissolved oxygen tension and mechanical forces on fungal morphology in submerged fermentation**. *Biotechnol Bioeng,* 1997, **57**:409-19.

AUTHOR INDEX